European Research on Sustainable Development

Carlo C. Jaeger • J. David Tàbara • Julia Jaeger
Editors

European Research on Sustainable Development

Volume 1: Transformative Science Approaches for Sustainability

Springer

Editors
Prof. Dr. Carlo C. Jaeger
Potsdam Institute for Climate Change
Impact Research
Transdisciplinary Concepts & Methods
Telegrafenberg A31
14473 Potsdam
Germany
carlo.jaeger@pik-potsdam.de

Dr. J. David Tàbara
Autonomous University of Barcelona
Institute of Environmental Science
and Technology
Campus UAB
08193 Cerdanyola del Valles
Spain
joandavid.tabara@uab.cat

Julia Jaeger
European Climate Forum
Postfach 60 06 48
14406 Potsdam
Germany
julia.jaeger.jaeger@googlemail.com

Softcover re-print of the Hardcover 1st edition 2011

e-ISBN 978-92-79-16583-2 e-Catalogue number KI-31-10-103-EN-N

DOI 10.2777/20753

ISBN, Catalogue number and DOI of the printed version will be printed on the backcover.

Published by Springer-Verlag Berlin Heidelberg 2011

ISBN 978-3-642-44499-9 ISBN 978-3-642-19202-9 (eBook)
DOI 10.1007/978-3-642-19202-9

Cover design: deblik, Berlin, Germany

Printed on acid-free paper

Springer is part of Springer Science+Business Media (www.springer.com)
DG Research (www.ec.europa.eu/research)

Foreword

Sustainable development is a challenge for scientific research in general and for European research in particular. It calls for radical transformations of the ways we do research and think about it, of the ways we shape research policies, and of the ways the scientific system interacts with the other components of society. As director for environmental research in the European Commission, I welcome this pair of books, that build on the Conference "Sustainable development: a challenge for European research",[1] that we organized in Brussels in May 2009. This event generated a beautiful positive energy and fostered an inclusive, but demanding, vision of sustainable development. Indeed, sustainable development is about vision, more than about prediction. Sustainable development is about confidence, more than about fear. Sustainable development is about equity and respect, more than about race and competition. Scientific research is neither a steady nor a stand-alone endeavour. The sustainability challenge calls for a web of constructive interactions, notably between science and policy-making, science and activism, science and philosophy, science and other forms of knowledge. I thank the editors of these books, and in particular Carlo Jaeger, chair of the scientific committee of the Conference, for putting together and publishing these books, and hence following up on the vivid conversation and the rich interactions initiated at the conference.

Manuela Soares

[1] http://ec.europa.eu/research/sd/conference/2009/index_en.cfm

Research for Sustainability and the European Union: From Wish to Will. A Manifesto

Europe and the Vision of Sustainable Development

Sustainability is a word full of promises, evoking peace, welfare, and harmony with nature. No wonder Europe has a strong wish for sustainability. Does it also have the will?

Europe's wish became explicit in 2001, and grew ever since to culminate in 2007. In January 2007, EU Commission President José Manuel Barroso stated: "Europe must lead the world into a new, or maybe one should say post-industrial revolution – the development of a low-carbon economy." In spring, the EU declared its "triple 20%" intention, setting ambitious quantitative goals for cutting greenhouse gas emissions, increasing renewable energy use, and improving energy efficiency. In summer, this step enabled the G8 summit of Heiligendamm to declare the aim to halve global CO_2 emissions by 2050. And at the end of the year, it was influential in keeping the momentum in the global climate policy process at the U.N. Climate Change Conference in Bali.

A year later, however, the biggest financial crisis since 1929 hit the world. That crisis made painfully clear how unsustainable the financial boom of the past decades had been. But the perspective of sustainable development has been largely absent in the haphazard way different European nations have tried to counter a global financial crisis that will shape the twenty-first century.

"Now you don't talk so loud, Europe" many voices say, from within as from without. These voices must be heard, because without a candid debate about the challenge of sustainability, Europe cannot develop the will to meet it. And it is the absence or presence of a will to meet shared challenges that makes political communities stumble or flourish.

If Europe is to develop this will, sustainability must stay on the top of its agenda, especially in difficult times. This will take a long and difficult journey of learning by doing. On one hand, it is clear that the current pattern of global development places a growing burden of huge risks on future generations – of nuclear war, of financial meltdown, of flooding the coastal cities of the world, to name but three of them. On the other hand, it is by no means clear how these risks can be reduced, step by step, starting from today's institutional, technological, and cultural conditions.

The Opportunity of Research for Sustainability

Europe cannot solve these problems alone, nor can it tell the rest of the world how to solve them. But it can make key contributions on the way to sustainable development. Against this backdrop, research for sustainability is a major opportunity for Europe to assume its responsibility as one partner among many in the conversation of humankind.

Here again we are faced with the needed transition from wish to will. The Seventh Research Framework Programme (FP7) has sustainability as one of its overarching goals. It offers plenty of opportunities for research in areas like health, energy, climate change, and food, and it encourages sustainability oriented research in fields as diverse as new production technologies and the humanities. Moreover, the European Research Council, founded as part of FP7 with the explicit aim of fostering "investigator driven frontier research", can nurture the next generation of scholars needed to advance research for sustainability. The newly founded European Institute for Innovation and Technology (EIT), based on highly integrated public-private networks of universities, research organisations and businesses, offers another critical opportunity to advance such research.

With such instruments at hand, we – European scholars confronting the sustainability challenge and being engaged through this very Manifesto in a conversation with the European Commission – call for opening a new chapter in the long history of European science and technology. This is the opportunity, but also the responsibility of European research today. There is a clear and present danger that large fractions of European research budgets will be spent – even in the name of sustainability – on a kind of research that cannot be expected to effectively address that challenge.

To transform the wish to foster sustainability through research into the will to do so, difficult questions must be pondered. Is research a matter of utility for given purposes, or is it a creative activity that transcends utilitarian reasoning? Is the fact that all European nations together spend much less on defence research than America alone, and that the current EU budget for defence research is only symbolic, something to be proud of, or is it to be changed by gradually building up a European military-industrial complex? What kind of research does sustainability require in the face of European agricultural policy? The answers to these questions are far from obvious, but if questions like these are not openly debated, the wish for sustainability will lead to wishful thinking, not to effective action.

A Passion for Quality

Research for sustainability can only grow if knowledge and action are intertwined in an on-going experience of learning by doing. In particular, learning from past mistakes and successes may be more fruitful in dealing with the challenge of sustainability than doomsday scenarios that leave no sense of choice. In the face of global risks, research must not sow paralysing fear, but anticipate possible changes and provide alternatives.

In this perspective, research for sustainability needs to overcome the mental frames that have blinded scholars and decisions-makers to the instability of the current world economy. The conceptual device of a representative rational agent has obscured the way interacting markets for commodities, land, and financial assets would lead to the financial crisis of 2008. In the future, the complex dynamics of socio-ecological systems – involving a multitude of heterogeneous agents embedded in shared environments – must become a key theme of research for sustainability.

The resulting research agenda will require much greater ingenuity, creativity, and patience than may appear at first sight. Research for sustainability needs skills of trans-disciplinary teamwork that are not part of traditional academic curricula. Sooner or later, European researchers should even be able to combine scholarly specialization with the philosophical literacy required to reflect on the relation between research, sustainability and the tradition of European thought.

To tackle the sustainability challenge, science must also invite and welcome plurality much more than it did in the past. A naïve belief in scientific progress will need to mature into styles of scientific research shaped by respect for different traditions of inquiry. Moreover, there is a need to catalyse critical dialogues across different domains of discourse by focussing on the sustainability challenge. New chairs, institutes, and curricula explicitly dedicated to sustainable development can inspire research in widely diverse fields of professional specialization. And together with public policy and civil society, the world of business enterprises and professional associations has a key role to play in developing practical solutions and far-reaching innovations in business models.

Research for sustainability is gradually becoming a solution provider for many of the most important problems humankind is faced with. To strengthen such research, the European Union needs specific funding mechanisms (to fund also relatively small, but long-term projects), dedicated review mechanisms (because sustainability research is goal-seeking, not simply goal-oriented); different incentive structures for career paths (credit for work on processes in analogy to the credit architects get for buildings and designs); flexibility and leadership.

Europe has the means, the duty, and the passion required to provide a global platform for such research.

This manifesto was elaborated in February 2009 by the Scientific Committee of the EU-conference: "Sustainable Development – A Challenge for European Research".

Contents

About the Authors

Line J. Barkved (MSc Hydrology) has been employed at the Norwegian Institute for Water Research (NIVA) since 2001, where she currently holds a permanent position as Research Scientist. Her work is mainly related to modelling of hydrology, nutrient losses and climate change. She has working experience from the Arctic, Europe, India and Vietnam. She has during her working years had a central role in the co-ordination team of several EU-projects, including the STRIVER project.

Dag Berge is senior research scientist at the Norwegian Institute for Water Research (NIVA). Originally educated as a limnologist, his main expertise is the eutrophication of lakes and reservoirs, phosphorus loading models and phytoplankton. During the last 15 years he has been working on EIA of hydropower development in Asia and Latin America, as well as water management planning, pollution abatement planning, monitoring of lakes and rivers, and drinking water supply. He has project experience spanning more than 20 countries.

Jos Bruggink obtained his masters degree in econometrics at the Erasmus University Rotterdam and his Ph.D. degree in development economics at the University of Oregon (USA). He started his professional career in energy at Resources for the Future in Washington DC. In the late 1970s he returned to the Netherlands as energy policy analyst for the Energy research Centre of the Netherlands (ECN), where he became unit manager of ECN Policy Studies in 1989. In 2004 he shifted back from management to research with a focus on long-term energy transitions, climate change policies and world energy markets and was appointed as a named professor at the Faculty of Earth and Life Sciences (Institute for Environmental Studies) of the Vrije Universiteit Amsterdam. He is author of Bruggink, J. J. C. 2005. *The next 50 years. Four European energy futures.* The Netherlands: ECN Publications.

Andrea Colantonio is Research Coordinator at LSE Cities, London School of Economics and Political Science, United Kingdom. He is an economist and urban geographer who specialises in sustainability and development studies in both developing and developed countries. Since graduating in economics at the

University of Rome, he has worked and researched in numerous international universities. In 2002, Dr. Andrea Colantonio also worked for the Environmental Industries Commission, London, participating in the Commission's Parliamentary activities. His main research interests are in the areas of sustainability (assessment methods, policy, planning and finance), institutional governance and economic development. He is main author of 'Urban Regeneration and Social Sustainability. Best Practice from European Cities (2010).

Johannes Deelstra holds a permanent position as senior researcher at Bioforsk (Norwegian Institute for Agricultural and Environmental Research). Deelstra studied Irrigation and Drainage at the Agricultural University of Wageningen, The Netherlands. He has a long time experience in issues related to the interaction between agriculture and the environment and has a special interest in the role hydrology plays in nutrient transport and erosion in small agricultural catchments. He has experience from working in Norway, the Baltic countries, Russia, Central and South America and Africa.

Nicole Dewandre is counsellor of stakeholder's issues in the Directorate General "Information society and media". She studied applied physics, engineering and economics at the University of Louvain, operations research at the University of California (Berkeley) and philosophy at the Free University of Brussels (ULB). She entered the European Commission in 1983, working as a scientific officer in the Directorate General for Research, in the field of prospective, linking societal concerns with technology policy until 1986. She became a member of the Central Advisory Group and then the Forward Study Unit, dealing with strategic analysis of research and industrial policy for the President of the Commission (1986–1992). In 1993, she supported the Belgian Presidency of the European Union in the areas of industry, energy, and consumer policies. Back in DG Research, she has been developing the "Women and Science" activities and she has contributed to the opening of the EU research policy to civil society, before being in charge of the "sustainable development" unit that has been put in place in DG Research between 2007 and 2010.

Niki Frantzeskaki MSc, has worked as a PhD researcher at Delft University of Technology, the Netherlands. She is currently working with DRIFT, Erasmus University Rotterdam, The Netherlands focusing on institutions and policy transitions. She earned an MSc in Engineering and Policy Analysis from Delft University of Technology, the Netherlands and an MSc/BSc in Environmental Engineering from Technical University of Crete, Greece. Her research interests include policy analysis, policy dynamics, social-ecological systems governance, institutions and transitions. Since her appointment in Erasmus University Rotterdam, she has been focusing on the intersection of institutions, policy transitions and dynamics for infrastructure systems (energy and water).

Geoffrey D. Gooch is Professor of Political Science at Linköpings University, Sweden, and at the IHP-HELP Centre for Water Law, Policy & Science, University of Dundee, Scotland. He is an expert in water and environmental policy analysis, public participation and communication, and he has conducted research in many parts of the world. He is currently coordinating the EU funded 'LiveDiverse' project which studies biodiversity and livelihoods in Costa Rica, Vietnam, South Africa and India.

Richard F.S. Gordon is Chief Executive of the New Zealand Crown Research Institute Landcare Research Manaaki Whenua. This government-owned organisation has 400 staff conducting science and innovation on challenges in the terrestrial environment: biodiversity, biosecurity, soils and land-use, greenhouse gases and carbon sinks, and the growth of organisations within environmental limits while meeting community and market requirements. Richard trained in Natural Sciences at Cambridge University, gaining a doctorate in 1981. He worked in agrochemical research and development in the UK and Japan before joining Landcare Research in 1995. He established the Institute's capability in Sustainable Business, led the preparation of New Zealand's first corporate sustainability report (for Landcare Research in 2000), and was a member of the first Stakeholder Council of the Global Reporting Initiative. He led the research programme, Building Capacity for Sustainable Development, and became CEO in May 2011.

Dr. Bruna Grizzetti is a researcher at the Joint Research Centre of the European Commission in the Institute for Environment and Sustainability. She has experience in hydrological and nutrient modelling at continental scale in support to the development and the implementation of environmental European policies. Her research interests involve interdisciplinary projects for sustainable water management in developing countries. Since 2007, Bruna Grizzetti is a member of the Coordination Team of the European Nitrogen Assessment, funded by the European Science Foundation.

Carlo C. Jaeger is Professor for modelling social systems at Potsdam University in Germany, chair of the research domain "Transdisciplinary Concepts and Methods" (TCM) at the Potsdam Institute for Climate Impact Research, and chair of the European Climate Forum. TCM develops climate impact research guided by stakeholder dialogues and using mathematics as a tool to meet conceptual challenges. He has been professor at the University of Darmstadt in Germany and head of the human ecology department at the Swiss federal institute for environmental science and technology. He is member of the Scientific and Technical Council of IRGC (International Risk Governance Council), and has served on the boards of various scientific organizations. He holds degrees in economics (Ph.D., Frankfurt University), sociology (diploma, University of Berne), and human ecology (habilitation ETH Zurich) and has worked extensively on interactions between technological progress and environmental problems, in particular the role of information technologies for urban development. He has also considerable research experience in the field of stakeholder dialogues. His current research interest is focused on the role of financial markets in managing climate change and is the coordinator of the EU project Global Systems Dynamics and Policies (www.gsdp.eu).

Jill Jäger is an independent scholar based in Vienna, Austria and a core member of the European Sustainability Science Group (www.essg.eu). Since 1979 she has worked as a consultant on energy, environment and climate for numerous national and international organizations. In September 1994 she joined the International Institute for Applied Systems Analysis (IIASA, Laxenburg) as Deputy Director for Programs, where she was responsible for the implementation and coordination of the research program. From October 1996 till May 1998 she was Deputy Director of IIASA. From April 1999 till October 2002, Dr. Jäger was Executive Director of the International Human Dimensions Programme on Global Environmental Change (IHDP: www.ihdp.org). Her main field of interest is in the linkages between science and policy in the development of responses to global environmental issues. Among her later books are: *Environment, Forced Migration and Social Vulnerability* (with T. Afifi Springer 2010), *and Our Planet: How much can Earth take?* (Hauss Publishing, 2009).

Derk Loorbach is a senior researcher and consultant at the Dutch Research Institute for Transitions (Drift) at the Erasmus University Rotterdam, where he received his Ph.D. in June 2007. Drift is a pioneering interdisciplinary institute that combines cutting edge research with close cooperation with policy and business to further sustainable development in practice. Central theme in Derks research is the development of the transition management approach as a new governance-model based on complex systems' thinking, governance theories and sociology. It aims to facilitate and direct processes of societal change in the direction of sustainability through transition arenas and experiments. Derk is currently involved in various transition arenas, innovation programs and envisioning practices in areas or urban development, energy, health-care and housing. His research is seen as example of 'sustainability science', combining fundamental with action research to contribute to sustainable development in practice.

S. Manasi is currently working as Assistant Professor in Centre for Ecological Economics and Natural Resources, Institute for Social and Economic Change, Bangalore, India. She has worked various areas of water resources – rural water supplies, urban water supplies and lately on integrated water resources management. Her areas of research are water resources, solid waste management and climate change.

Mary Midgley is a professional philosopher whose special interests are the relations between humans and the rest of nature, and the troubled frontier between science and religion (particularly in cases where science becomes a religion). Born in 1919 she took her university degree in Classics, Philosophy and Ancient History at Oxford during the war. After graduate work, she lectured in Philosophy, first at the University of Reading, then at that of Newcastle on Tyne, where she became a Senior Lecturer. She still lives in Newcastle. She is the author of many books including **Midley, M.** 2010. *Science And Poetry, The Ethical Primate, Science As Salvation, Utopias, Dolphins and Computers and Animals And Why They Matter* for which she was awarded in 1986, jointly with Konrad Lorenz, the Premio Gambrinus Giuseppe Mazzotti. Her memoir, *The Owl of Minerva*, was published by Routledge in 2005.

Dr. Udaya Sekhar Nagothu (Research scientist) holds a Ph.D. in Development Studies, Agriculture University of Norway (1999). He has considerable research experience in (i) institutional and policy analysis, (ii) stakeholder analysis and stakeholder participation, (iii) conflict analysis and conflict management, (iv) tenure and property rights and (v) integrated watershed management. He participated in a project on Triangular institutional cooperation programme (Ethiopia-India-Norway). Other countries worked in are Vietnam, India and eastern Sri Lanka.

Dr. Ingrid Nesheim is a researcher at the Centre for Development and the Environment, University of Oslo. Nesheim is qualified in the field of plant ecology, ethnobotany and multivariate statistics. She has long experience with interdisciplinary research in various research projects involving fieldwork from rural areas in the tropics, and her perspective encompasses both environmental and social science. Nesheim teaches Interdisciplinary methodology at master level at the University of Oslo.

Dr. Dang Thi Kim Nhung a researcher at Institute of Geography-Vietnam Academy of Science and Technology is a geographer with a PhD in Geography from the IoG Poland Scientific Academy (1988–1991). She has wide experience in EIA and environmental management, and has participated in many international projects funded by ASEAN and the EU.

Eleonore Pauwels is a visiting scholar with the Foresight and Governance Project at the Woodrow Wilson International Center for Scholars in Washington DC. Her primary focus is a comparative and critical analysis of the EU and US approaches towards the societal governance of synthetic biology. She is also examining the challenges that new forms of biotechnology pose for political and public policy organizations, and the regulatory innovations that emerge alongside developments in cutting edge technologies. Her past research has included risk communication, citizen participation in technical controversies, and the epistemology of regulatory and social scientific experiments. Eleonore holds a M.A. in European Public Policy and Politics awarded by *L'Institut d'études politiques de Paris* (IEP) and Bath University within the European American University Consortium. In 2006–2007, she was part of the Governance & Ethics Unit of the Directorate-General for Research at the European Commission.

Antonio Lo Porto is an ecologist with a graduate degree in Natural Sciences from University of Bari. He is researcher at the IRSA-CNR (Dept. of Water Resources Management) where he focuses on the integrated use of GIS, simulation models and statistical methods to study diffuse water pollution from agricultural sources and to design management plans for river basins. He also develops techniques for planning and designing measures to mitigate impacts on receiving water bodies.

Dr. Santiago Beguería Portugues is Postdoctoral Fellow and has a PhD in Geography from the University of Zaragoza. He has published in the main international journals in hydrology. His main works are related to rainfall-runoff relationships, extreme events and the location of sediment sources. He is an expert in GIS and remote sensing.

Dr. Alistair Rieu-Clarke (LLB, LLM, PhD) is a Senior Lecturer at the IHP-HELP Centre for Water Law, Policy & Science, University of Dundee. Alistair is actively involved in a range of multi-disciplinary, international research collaborations largely centering around measuring the effectiveness of (trans-boundary water) governance regimes and designing participatory research methods in the context of water resources management. Additionally, Alistair runs a postgraduate module on International Law of Water Resources, as part of the UNESCO Centre's Water Law: Water Leaders Programme, and supervises a number of PhDs.

Vandana Shiva is trained as a Physicist and did her Ph.D. on the subject "Hidden Variables and Non-locality in Quantum Theory" from the University of Western Ontario in Canada. She later shifted to inter-disciplinary research in science, technology and environmental policy, which she carried out at the Indian Institute of Science and the Indian Institute of Management in Bangalore, India. Dr. Shiva is one of the world's most renowned environmentalist. Time Magazine identified Dr. Shiva as an environmental "hero" in 2003 and Asia Week has called her one of the five most powerful communicators of Asia. Dr. Shiva has pioneered the organic movement in India and established Navdanya, the country's biggest network of seed keepers and organic producers. Dr. Shiva has authored many books including *Soil Not Oil, Earth Democracy, Stolen Harvest, Staying Alive, Water Wars* and *Biopiracy*. Among her many awards are the Alternative Nobel Prize (Right Livelihood Award, 1993), Order of the Golden Ark, Global 500 Award of UN and Earth Day International Award. Lennon ONO grant for peace award by Yoko Ono and Honourable Mayor of Reykjavik. Dr. Shiva also serves on the boards of many organizations including World Future Council, International Forum on Globalisation and Slow Food International.

Per Stålnacke (Ph.D.) is Head of Department and a Senior Research Scientist at Bioforsk (Norwegian Institute for Agricultural and Environmental Research). He has interdisciplinary Ph.D. degree in Water and Environmental Studies from

Linköpings University, Sweden. Stålnacke has long-term experience in issues devoted to integrated water resources management with particular emphasis on studies of pollutant fluxes in river basins and statistical analysis of historical environmental monitoring data. He has edited 3 books in the field of integrated water resources management. Stålnacke has a scientific interest to bridge the gaps between research and practical water management.

J. David Tàbara is a senior researcher, professor qualified, at the Institute of Environmental Science and Technology of the Autonomous University of Barcelona (UAB). He holds degrees in economics, environmental economics, business administration and developed his Ph.d. in the emerging field of environmental sociology (1995). At UAB he lectures social science method for Integrated Assessment and Sustainability Science. He has 15 years experience of EU research on sustainability issues. He has published extensively on socio-environmental theory and integrative environmental methods, with particular attention to questions of public perception, social learning, communication and public participation for sustainability. In 2009 he received Research Excellence Award at UAB. Recently, he developed the appraisal method on the 'climate learning ladder' (*Environmental Policy and Governance*, 2010, vol. 20:1–11).

Haakon Thaulow is Senior Enviromental Advisor at the Norwegian Institute for Water Research (NIVA). He has wide experience in Water Managment from the government sector (Ministry of Enviroment and State Pollution Control Authority) and from applied Water Research at NIVA. He was for 15 years (1985–2000) Managing Director at NIVA. He was in charge of the Norwegian Master Plan for Water Resources (1980–1984), Chairman of Norways Water Resources Commission (1988–1992) and President of the European Water Association (2003–2005).

Wil Thissen is a Professor of Systems and Policy Analysis and leader of the Multi-Actor Systems research programme at Delft University of Technology. He has over 25 years of experience working at the interface of systems analysis and modeling on the one hand, and policy and decision making at the other, and is one of the founders of innovative interdisciplinary educational programs in Technology, Policy and Management at TU Delft. His current research focuses on energy transitions in complex socio-technical systems, multi-actor systems modeling, and dealing with (deep) uncertainties in strategic decision making, with a particular interest in applications to energy systems, water management, and adaptation to Climate Change.

Paul M. Weaver Economic geographer and sustainability science practitioner. Experienced research manager and has been a Principal Investigator for several EU and national research projects, including MATISSE, AIRP-SD. Thirty years experience in field of environmental resource management and sustainable development. Advisor to the 1992 UNCED process and has worked as a consultant for international organizations, universities and companies. He specialises in sustainability assessment, using place-based and chain-based approaches, and the use of evaluation as a tool to drive innovation and to upscale from successful niches in applied transition work. He co-edited with J. Jager and J. Rotmans (2008) the special issue of the *International Journal of Innovation and Sustainable Development* on Integrated Sustainability Assessment, (Vol. 3, Nos. 1/2).

Introduction

J. David Tàbara

This book brings together a unique cohort of scholars and practitioners who examine the practical meaning of sustainability for science and for policy in interaction with wider publics in the face of global change. The contributions that follow address the transformations required in mental framings, institutional settings, and research practices to support the vast societal transition towards sustainable development. The interdisciplinary background of the authors means that the concepts, arguments, and ideas advanced in this volume are not based on one single approach or scientific perspective, but draw from a wide variety of personal and professional experiences. Most of the chapters in this book are built on the lessons learnt from major international research efforts on environmental issues and sustainability, in areas like community-based resource conservation, climate change, water, or land planning. Because of the pragmatic orientation of this book, which focuses on articulating an in-depth reflection for the implementation of concrete changes and decisions, we have explicitly avoided excessive academism. Our main goal is to contribute to the discussion for a practical and positive vision of science, policy, and public interactions in the face of global environmental change and unsustainable development.

This volume is based on a selection of key papers given at a conference organised by the EU DG-Research in May 2009 on 'Sustainable Development. A Challenge for European Research'.[1] The first three chapters deal with the concept of sustainable development and sustainability research from quite different but closely interrelated positions: philosophy, environmental advocacy, and policy practice. These are then followed by the seven awarded papers at this conference plus two final contributions by Paul Weaver and Jill Jäger who analyse the potential contributions and evolution of sustainability science and its possible futures. In the concluding remarks, Carlo Jaeger and J. David Tàbara look at the role of new forms

[1] http://ec.europa.eu/research/sd/conference/2009/index_en.cfm.

J.D. Tàbara (✉)
Institute of Environmental Science and Technology, Autonomous University of Barcelona, Campus UAB. E-08193, Cerdanyola del Vallès, Barcelona, Catalonia, Spain
e-mail: joandavid.tabara@uab.cat

C.C. Jaeger et al. (eds.), *European Research on Sustainable Development*,
DOI 10.1007/978-3-642-19202-9_1, © European Union, 2011
Published by Springer-Verlag Berlin Heidelberg 2011

of making science for sustainability, as well as its requirements and potential, against the background of the crisis of Western science.

Our journey starts with a powerful critique by Mary Midgley of the current and dominant notion of 'development', and how the idea of sustainability has helped to unfold some of the contradictions and myths embedded in it, such as the denial of resource limits. To Midgley, our current understanding of 'development' still reflects the colonial thinking ingrained in the particular belief of progress as a linear and never-ending process, whereby less mechanised societies inexorably move forward to achieve the ultimate state represented by the model of the metropolis: 'Belief in progress seemed to mean that a fixed course of life was set before all peoples, a necessary journey away from a primitive state, a single racecourse on which all were travelling, some much faster than others, towards increased use of machines.. It is perhaps strange that we still use this imagery today when we talk of developing and developed nations (...). Talk of development implies a pre-set course of life, like that by which tadpoles become frogs and caterpillars become butterflies. So our current use of it means that we're all going the same way, only some nations are ahead of others. This thought has often been expressed by describing less mechanized people as essentially children – beings who are doubtless worthy but are behind us in the journey of life (...) It has actually been devised as a euphemism, a polite substitute for cruder words like primitive, backward or uncivilized'. The problem with this world-picture is that many communities in the World along with a growing number of signals sent by the Earth, including its biophysical limits, do not seem to be cooperating with it. The support required for this preset route in terms of its social and ecological conditions is fading. In the face of this situation, Midgley calls for a new type of extended humanism and of science which does not become 'species-conceit', which is less technophile, and which demands an enormous mental and cultural shift. A new *vision* of development and of life on Earth is required, one which corrects these ideological and practical flaws, and which is based on new insights as well as on facts. In her view, Gaian thinking could help to overcome the prevalent atomistic views in science and to support the transformation of the current collective attitudes which negatively affect the health of the planet.

In line with these reflections, Vandana Shiva places equity at the centre of the challenge for sustainability: 'without equity, sustainability will definitively go wrong (...) it is only when you are organised as a community that you can protect the forest (...) that you can do water harvesting so that the water will be shared equitably. You can exploit resources under unequal regimes. You cannot conserve resources under unequal regimes. It is an asymmetric relationship'. Shiva defends that linear thinking needs to be turned into circular thinking, and that we need to avoid the 'monocultures of the mind' created by the 'commoditification' of the land and natural resources. Equity in sustainability entails the recognition that we all members of the Earth community, both human and non-human beings, have the right to ecological space, and when certain activities undermine this natural right, as happens with commerce, such activities must stop. For her, humanity is in a very critical moment: 'we are amidst a democratic transition from an oil-centred, capital intensive system of production, to a soil-centred, people-centred system of production (. . .). We will either have justice, sustainability and peace together, or we will descend into ecological catastrophe social chaos and conflict'.

Well aware of the above debates, Nicole Dewandre places sustainability firmly on the agenda of political action to shape an open future – rather than indulging in catastrophism or denial. Neither catastrophism nor denial can provide a positive transformative vision which is precisely what sustainability is all about: 'sustainability is what lies between catastrophism and denial. The challenge is to fill this concept with a positive meaning, reclaiming a collective sense of purpose'. This requires 'reconnecting knowledge with confidence in human action and in the future' starting with what there is here and now but connecting it with a sense of continuity with future generations –an undertaking which ultimately depends on 'our capacity to decide if we love this world enough to let them undertake something we did not predict, and even not foresee'.

Andrea Colantonio looks at new emerging themes in the discourses on sustainability, by reviewing the concept of 'social sustainability' and how this concept has been translated into policy. Social sustainability relates to activities and issues which have to do with 'how individuals, societies and communities live with each other and set out to achieve the objectives of development models'. It 'blends traditional social policy areas and principles such as equity and health, with issues concerning participation, needs, social capital, the economy, the environment, and more recently, with the notions of happiness, well being and quality of life'. However, Colantonio notices that the number of dimensions that fall into the umbrella of this concept are growing – making its operational use in policy more complicated: 'The chronological review of these themes suggests that traditional themes, such as equity, poverty reduction and livelihood, are increasingly been complemented or replaced by more intangible and less measurable concepts such as identity, sense of place, and the benefits of "social networks"'. This is similar to what is happening to Impact Assessment and Strategic Impact Assessment which are evolving in a way that encompass a growing number of issues leading to the emergence of 'Sustainability Assessment' – which in turn also derives a new set of indicators and metrics.

Jos Bruggink, takes up again the theme of catastrophism under the idea of 'doom' and how this relates to sustainability action. By distinguishing between three main scientific perspectives of sustainability – the normative-prescriptive, the analytical-empirical, and the strategic action-oriented-, his analysis focuses on what propositions could be best provided to support the latter. First, policy makers aiming at taking a proactive attitude to sustainability need to balance the risk-taking position, rather optimistic, based on the assumption that the world is manageable, with that which tends to retreat and to take a rather fatalist position about the possibilities for any world transformation (referred to as 'shaping' versus 'hedging' strategies). Second, Bruggink calls for a reframing of the usual characterisation of the 'three corners' of energy doom, that is, energy scarcity, climate disasters and energy poverty. He suggests that it would be better to look at these issues in a hierarchical way, locating global inequality at the top of challenges, followed by environmental goals such climate change, and economic ones (profits). And third, it would be necessary to link these three corners of energy doom in a dynamic way, instead of dealing with them in isolation as it is often the case: 'climate change mitigation policies are likely to become effective if and only if energy scarcity problems become more urgent and visible, and that persistent global energy poverty can be addressed more effectively if and only if climate change problems become more urgent and visible'.

Derk Loorbach, Niki Frantzeskaki and Wil Thissen describe and adopt a transitions approach as an integrated perspective to research and to governing sustainability transitions. In their view, despite the uncertainties and the complexity of the issues that need to be tackled, sustainability oriented transitions can be steered and managed although this requires of the active involvement of scientists, policy makers, and various stakeholders both in the development of new knowledge and in its application. Four propositions are put forward in this endeavour: (1) emphasis on long-term systems learning processes, (2) focus on radical change through incremental steps 'to erode' and 'ultimately dismantle' the existing structure, (3) creating spaces for experimentation in which visionaries and front-runners can develop their niche innovations and (4) to securing sustainability values that address long-term orientation and intergenerational justice. As transition management aims at fundamental shifts in social structures and practices, it is a 'research topic that by definition cannot be developed in a traditional, purely scientific sense'.

J. David Tàbara, develops these ideas further by focusing on the institutional requirements for an 'Integrated Climate Governance' (ICG). The novel concept of ICG results from bringing together several strands of thought relevant for sustainability and climate policy, including Integrated Assessment, complex social-ecological systems science, and multi-level governance. It is also based on the notion of transition management, insofar that it goes beyond the idea of 'assessment' and it is based on *appraising* both climate risks and opportunities to support individual transformation and action for sustainable development. Integrated Climate Governance provides a unique opportunity for global cooperation and development, as well for the advance of sustainability science and innovation. However, progress can only be achieved if the new tools, methods and processes for ICG are built on integrated empirical experiences that consider local and regional needs and ambitions and fully take into account processes occurring at other levels of governance and policy domains. It is argued that ICG is mostly an institutional challenge that lies in the intersection between the assessment of risks and opportunities, the engagement with and empowerment of the public, and the development and implementation of policy instruments.

Eleonore Pauwels provides a critical analysis to explore possible contributions of Science and Technology Studies (STS) to sustainability research by focusing on the case of emerging technologies, such as synthetic biology. In her view, 'the public and policy debates surrounding synthetic biology have been narrowly focused around a utilitarian calculation of its technological benefits versus its potential regulatory risks'. STS can contribute to the assessment of potential risks and benefits of this new wave of techno-engineering integration insofar as we understand that 'the ultimate meaning and implications of sustainability as a policy issue are, for the most part, not intrinsic but a human construction'. Pauwels further elaborates on this position by taking a Latourian perspective whereby synthetic biology objects can be understood as hybrids: human-nature co-productions (and risks) which are neither only human nor natural but are both, and of which their intertwined dynamics increasingly determine our global social-ecological practices and institutions. Such hybrids tend to amplify and often increase the unexpected effects and unpredictable behaviour of both humans and non-humans in ways that require new approaches to thinking and to governance organisation. Therefore, for sustainability science, the challenge is to develop the appropriate institutional processes capable to yield and govern alternative plausible pathways open to public scrutiny so as to avoid the most irreversible unwanted situations.

Richard F. S. Gordon, reflects on the lessons learnt from a series of sustainability and innovation research projects and scenario exercises carried out by Landcare Research in New Zealand, a country which can serve as a 'laboratory for solutions for other countries'. Of particular interest are their experiences to integrate Māori values and practices into developing national research agendas. Among other issues, Gordon's contribution makes us consider the importance of the rights of indigenous peoples and how the violation of these can be compensated. However, what is relevant in this context is the need to examine the diversity of elements and criteria that such cultures employ to organise their business and collective activities in the contemporary world, so that they can be actively used to inform innovative research agendas for sustainable development at the national level. Thus efforts should not only be made to protect diversity because of its intrinsic value, but also because it can also be very valuable to support sustainable development – and in turn, by making the best use of its potential, appreciate its true value to protect it. In the same way as 'impacts in complex systems come from influencing paradigms and mechanisms of governance' the integration of diversity can also help to modify and innovate governance structures that drive different patterns of development.

Per Stålnacke and colleagues examine the concept of Integrated Water Resources Management (IWRM) and how this concept can be used to promoting sustainable water management. To this end, they summarise the main findings from a large EU project carried out in four river basins of the world: Glomma (Norway), Tagus (Spain and Portugal), Tungabhadra (India) and Sesan (Vietnam and Cambodia). They provide a show-case on how particular methodologies and tools, such as the environmental flows, pollutions models, actor-network analysis, or scenarios can play a special role to further implement IWRM strategies in the context of sustainable development. However, this also requires the creation of assessment capacities that take into account several science perspectives and a regulatory framework that includes governance structures, law, land use changes and other elements. Among these, stakeholder participation – especially that which occurs in the interface between science, policy and civic society organisations – is one of the most critical elements for its practical implementation.

Paul Weaver starts his discourse by acknowledging the 'paradox' of sustainable development and by placing his contribution in the discussion between 'purist' versus 'pragmatic' approaches to sustainability. How humankind is going to manage critical global shared pool resources such as climate and biodiversity is used as an illustration of applied problems for which we need to develop pragmatic responses. Purist stances stress the importance of a particular approach as being the 'key' for sustainability. These are often defined along the lines of single 'structural themes', such as policies, markets, or technologies, and advocate (typically) whole-scale reforms to prevailing conditions on the concerned theme. Purism reflects particular stances in respect to the 'how' question of what we might do in a consistent logic to chart a future pathway. Pragmatism, by contrast, stresses that change processes in complex systems are multi-level, multi-dimensional, and multi-speed and that innovations developed at lower levels of scale need to be adapted both to local contexts and to higher-level framing conditions and dynamics. Pragmatism suggests that sustainability at the global level depends on achieving a patchwork of sustainable solutions at lower scale levels, each developed 'in context' and adapted to local needs and conditions. In turn this stresses the need for combinations of mutually supportive and reinforcing top-down and bottom-up innovations that engage synergies among innovations on many different fronts. The challenge of implementing a strategy based around a patchwork of context-sensitive local solutions implies new, additional and different roles for science and scientists in facilitating mediation and modulation between top-down and bottom-up innovations. Especially it suggests a key role for scientists in supporting processes of community-based governance of ecosystems and environmental threats and in helping communities to develop solutions that take advantage of opportunities in the dynamics of framing conditions as these arise. These are roles that go even beyond applied problem solving, since they involve nurturing mobility in forms of governance and in social values. These are not easy roles for scientists to play or ones that fit well with prevailing scientific cultures and prevailing organisational and institutional arrangements in science. However, it is through these roles – played out locally in situations and contexts throughout the world over the coming years – that scientists will probably have most impact on the prospects for more sustainable development.

Finally, Jill Jäger looks at the experiences on sustainability science in Europe and reflects on the contributions of the EU DGXII 2009 Conference. Jäger argues that sustainability science could contribute to the development of measures which deal with persistent problems of unsustainability by the design of structured processes of dialogue between relevant stakeholders to support experimentation and learning for sustainability. Nevertheless, and to become relevant, major changes in the organisation and funding of science are required, and in particular, in the way research proposals are evaluated and reviewed. In effect, sustainability science is a goal-searching rather than a goal-driven endeavour. The success of a project cannot be assessed depending on a specific objective predetermined *ex ante* –mostly because both the goals and the means are subject to large uncertainties and ultimately will have to be uncovered by the actual process of research. Being necessarily open-ended, iterative, participatory, and with a focus on understanding and transforming the complex interactions between social and natural systems, sustainability science faces large barriers for its institutionalisation, but also growing opportunities, as demonstrated by the fast growth of networks and publications not only in Europe, but also worldwide in this transdisciplinary field.

From the contributions of this book it seems apparent that two broad possible futures or understandings about what research for sustainable development means are possible. On the one hand, a regime-based 'research for sustainable development', could be centred on generating new products and services and reducing the relative impact of certain production and consumption processes through ecoefficiency. Under this approach no explicit purpose may be made by scientists to actively engage in developing alternative visions of the future, in transforming institutions, or in modifying the organisational patterns of behaviour and cultural assumptions that ultimate drive the unsustainability of current development. While such an approach is likely to be welcomed by the incumbent regime, it is also likely to be insufficient to deal with the increasingly interconnection, intensity and scale of the new global risks and unsustainability challenges that the present world is to confront now and in coming generations.

On the other hand, an alternative approach, as the one defended in this book, may take a more transformative, process-based, integrative and multi-scale orientation. Current science seems to work well for problems which are compartmentalised, but does not perform well to provide answers to problems that are systemic, interdependent and multi-faceted and in which changes in values and institutions need to be considered. Our evidence shows that there is growing attention paid by funding agencies and universities for this new type of reasoning and practice. Diversity in science and innovation is always welcome, so even if it were only for precautionary reasons, it would be foolish to dismiss the idea and the possibility that science for sustainability can be done in different ways. Our point is that many 'other' forms of transformative science for sustainable development are not only possible but are urgently needed. Transformative science for sustainable development is a requirement for the democratisation of knowledge in an increasingly open global system where the views of the final recipients of scientific and technological innovations – and their negative side effects in particular locations – require a much greater attention.

Sustainability science can contribute to sustainability learning. This means to learn not only about the possible outcomes – and the unintended negative consequences of human action – but also about the possible processes to design alternative pathways of development that are needed to cope with them. As mentioned by Mary Midgley, this calls for reconstructing human meaning within the boundaries of our social-ecological systems, taking into account all the diversity and interconnectivity of life forms and allowing for different patterns of organisation. Looking at development as a single linear, implacable, and ascendant process is not only a rather odd belief, but also a very dangerous one. New concepts, procedures and overall perspectives based on new ethical values and worldviews. Which in turn, also need to be researched and ultimately put into practice. In this respect, a main contribution of sustainability science could be helping to identify alternative pathways for development in which specific technologies, institutional arrangements and social-ecological patterns of organisation can be placed, according to visions, knowledge representations and experiences placed in concrete contexts democratically reflected. That is, instead of new technologies creating new path dependencies which decide, in a rather autonomous or unaccounted fashion what futures we have to live, interdisciplinary and transdisciplinary approaches based on alternative visions of the future could be devised to harness such innovations and avoid the more potentially perverse and negative irreversible effects of their unprecautionary implementation.

In sum, the works presented in this volume point out that new forms of transformative learning and researching are needed to address the multiple challenges of sustainable development. A new vision of knowledge and of the ways of engaging and carrying research is required, based on a diversity of perspectives, languages, and research practices – as well as a on a broader worldview of humans on Earth. Key challenges remain in the development of adequate institutions and capacities for people to work in the interface; for sustainability learning is likely to emerge not so much from innovations and developments occurring within the remits of science, of policy, or of public advocacy, but from the interaction between them. This book provides the handful of key insights on the possibilities developing alternative science approaches, as well as concepts and ideas necessary to reframe the role of science and of technology in the common quest for sustainability. Albeit the difficulties, ambiguities and barriers that sustainability scientists confront at the present, our approach intends, as pointed by Nichole Dewandre, to navigate between denial or catastrophism – while avoiding both – to provide a vision of what is feasible and desirable to transform science and its interaction with policy and the publics to contribute to sustainable development.

Developmental Doubts

Mary Midgley

1 Development and the Dream of Progress

On the face of things, the idea of sustainable development has something paradoxical about it, yet I think it has been very useful.

It is paradoxical because of the way in which we have habitually thought of development. If development meant simply spreading the equipment of Western civilization everywhere – if it meant providing all humans with tin cans, motorways, frozen meals and flush toilets – then it certainly couldn't be made sustainable. Even if it was a desirable aim, we couldn't do it. There aren't the resources for it.

All the same, the idea that some elements of this civilization can be used to spread lasting benefits has been really helpful. When the Brundtland Report first used this language, it managed to dispel a certain mist of unreality which had surrounded earlier efforts to draw attention to the state of the planet. Till then, practical administrators had often seen talk of environmental danger as unrealistic sentiment, something not necessarily wrong but like religion, best kept for Sundays. This whole topic called for such a long perspective that these people often couldn't see it as practical at all. Talk about it seemed to them just to express a mindless general objection to technology. By contrast, the language of sustainable development allowed them to keep their general ideal of development – the value of technology and the need to share it with less mechanised parts of the world. It only asked them to distinguish between more and less destructive ways of doing this. It made them start to ask questions, for the first time, about the long-term biological effects of colonial and foreign policies. Thus it cracked the shell of total denial.

But it by no means cleared denial out of our lives. All of us – even the most enlightened – are still mired in it to some extent today and we need to understand it better so as to grasp where it gets its strength. The trouble is that our beliefs are

M. Midgley (✉)
Independent scholar, Newcastle, United Kingdom
e-mail: dr.mary.midgley@googlemail.com

C.C. Jaeger et al. (eds.), *European Research on Sustainable Development*,
DOI 10.1007/978-3-642-19202-9_2, © European Union, 2011
Published by Springer-Verlag Berlin Heidelberg 2011

never just straight representations of immediate facts. What shapes them is the background visions that inspire us, the colourful, sweeping world-views that we are used to and that we take for granted.

Those simple views tend to have more power over us than the detailed, fiddly facts. And what powers our denial today is the splendid vision of Progress which has been growing more central in our culture for the last 300 years. All through the Enlightenment, the bold hope of future improvements on earth has been steadily replacing the religious belief in Heaven, as the influence of Christianity grew gradually weaker. People started to see the impressive March of technology, which distinguishes our civilization, not just as something useful but as a dominant symbol of enlightened living. It gave them a sense of control, an impression of reining in any forces that might threaten us.

2 Can the Sleeper Wake?

This deep confidence in technology naturally brought with it a poor opinion of less-mechanized cultures. Belief in progress seemed to mean that a fixed course of life was set before all peoples, a necessary journey away from a primitive state, a single racecourse on which all were travelling, some much faster than others, towards increased use of machines. It is perhaps strange that we still use this imagery today when we talk of developing and developed nations. The word develop, like evolve, originally describes the unrolling of a scroll or the opening of a bud – the revealing of something latent that was already fixed and predestined. Talk of development implies a pre-set course of life, like that by which tadpoles become frogs and caterpillars become butterflies. So our current use of it means that we're all going the same way, only some nations are ahead of others.

This thought has often been expressed by describing less mechanized people as essentially children – beings who are doubtless worthy but are behind us in the journey of life, Thus Auguste Comte described religion as a phase belonging to the childhood of our species – something that adult humans, who could use science instead, had outgrown. And colonisers clearly often did see those they colonised is childish in this way. Kipling expressed that vision somewhat alarmingly when he wrote of their mission-

> To wait in heavy harness
> On fluttered folk and wild –
> Your new-caught sullen peoples,
> Half-devil and half child -
> ('The White Man's Burden')...

Western colonisers are, of course, not the only group in the world who have used offensive language to describe cultures that they don't understand. This is actually quite a widespread habit. The reason why it matters so much now is because of what it reveals about our deeper attitudes. Kipling did not mean to be offensive; he was just thinking as the people around him thought – and indeed as those who administer empires usually have thought, even when their subjects were not foreigners. And today's current talk of developed and developing nations is certainly not meant to be offensive either It has actually been devised as a euphemism, a polite substitute for cruder words like primitive, backward or uncivilized. It just expresses our underlying myth – our quite genuine faith in the power of technology to provide all humans with a future that is steadily improving, perhaps indefinitely and for ever.

Such faiths are often harmless. But the unlucky thing about this one is that, just when we have communicated it to many of the cultures around us – just when we have converted them to technophilia and persuaded them to change their ways of life to suit it – we find that the world's climate is changing in a way that puts this possibility right out of date. This calls for a really enormous reversal of background thinking. Perhaps it is the biggest change required in human thought since the invention of agriculture. That is a change that would be hard to navigate at any time. But there is something in our own recent intellectual history which makes it specially hard for us even to consider.

3 Euphoric Humanism

The trouble here is the quite recent narrowing of our world-picture. Since the Enlightenment, we have steadily and deliberately refocused our vision to show humanity, on its own, as completely self-sufficient. We have played down all fear of God and of non-human natural forces which might limit our range. We have cultivated a kind of humanism which easily becomes species-conceit, – even extending sometimes to the idea that selected human activities constitute the aim of the whole cosmic process. In these scenarios humans appear as isolated from all other organisms, these being mere subordinates or opponents to be subdued. Thus the cosmologists John D.Barrow and Frank J.Tipler sketch a vision of the cosmic future in which our culture will send out intelligent machines that will gradually colonise the whole universe, finally reaching an Omega Point at which, as they put it, 'life will have gained control of all matter and forces, not only in a single universe but in all universe whose existence is logically possible,... and will have stored an infinite amount of information.... And this is the end'.[1] A footnote comments, 'A modern-day theologian might wish to say that the totality of life at the Omega Point is omnipotent, omnipresent and omniscient!'- i.e. it is God. It is interesting to note that, at this point, the 'totality of life' is supposed to consist solely of a batch of computers. Thus in this dream, progress, embodying simply the human academic's typical desire for power and information, is seen as supplying the meaning of the whole universe.

---

[1]Barrow, John D and Tipler, Frank L, The Anthropic Cosmological Principle, Oxford University Press 1986, pp. 677 and 682.

Barrow and Tipler are of course somewhat extreme. But the general force of such humanolatrous ideas is hard to exaggerate. For instance, as late as the 1900s quite enlightened sages such as William James and Freud preached the idea of a 'war against nature', a war which the human race was called on to wage and win. James wrote an essay called 'The Moral Equivalent of War', in which he noted the extraordinary enthusiasm that ordinary wars tend to arouse, and asked whether that fervour could somehow be redirected to less destructive channels. He wanted a displacement activity, a punch-ball, something like the ways in which baboons hit harmless passers-by and humans break china when they can't reach the real object of their wrath. He suggested campaigns which seemed to him quite harmless, for instance mining work, It clearly had not struck him that mining itself might be destructive. And, more generally, the idea that 'nature' might not actually be a punch-ball but a vulnerable system on which humans were totally dependent clearly never occurred to him. Freud argued similarly in Civilization And Its Discontents.[2]

Besides this difficulty about making a punch-ball of nature, however, there was also a problem about identifying what the interest of the whole human race actually involves. At first, during the age of colonization, human interests were identified quite simply with the interests of the colonizing elite. Thus Carlyle complained indignantly about the 'laziness' of black slaves in Jamaica because they didn't want to work as hard as the Europeans tried to make them. More lately, of course, this attitude has officially been dropped along with the rest of colonial thinking. Development is now supposed to mean that each region follows the course that best suits it, making the changes that are needed by its own people. But laissez-faire economics, which has become closely linked to the idea of progress, has given this system a somewhat peculiar twist. Thus, if an African country can make money by growing cut flowers for the European market, it may be urged to do that, even if this uses up land and water which its people need to grow their own food, or which is essential for its natural vegetation.

4 What Is It That Develops?

This kind of approach throws an interesting and unexpected light on the meaning of the term 'develop'. Is the organism which we now think of as 'developing' perhaps not so much a particular country as the economy of that country, or indeed the global economy, a strange, vast animal whose life-blood is the profits that flow in it, an animal which – like a cancer – always grows and can never contract? Indeed the meaning of 'development' here perhaps comes close to that of growth – another biological metaphor where an abstract entity is supposed to prosper in a way that has little relation to the fortunes of the people involved. Within that deliberately

[2]Published with The Future of an Illusion in 1927.

narrowed world-picture, with the strong economic twist that has marked it for the last century, this way of thinking can seem perfectly natural.

It's becoming clear, however, even to people who are very surprised about it, that this world-picture is rather like a dream in which the dreamer is disturbed by a faint recurrent tap – tap – tap – which eventually forces him to wake, and which turns out to be the noise of rain coming through the roof. On two fronts reality is breaking in. On one side, the people who are supposed to be being developed have begun to raise questions, refusing to accept the economic goals that others prescribe for them. And on the other – what is even more threatening – the physical state of the world is refusing to co-operate with our prescribed pattern of continuous improvement. It turns out that natural resources are limited and that climate change – ignoring the predicted economic and political pattern – is moving the other way, and doing it too fast to be ignored. In fact, both the two kinds of background support which were expected to fuel the endless March of Western progress – the social and the ecological basis -are fading away. The question is, can we find a way of thinking that enables us to do without that prospect?

5 Noticing the Planet

There are two big psychological and philosophical difficulties here. One of them concerns the unit of change. We have been accustomed to thinking of the human race in isolation – indeed for many of us in the West this has meant thinking of our own culture in isolation. The image of Man with which we grew up has represented him (repeat him) as a user standing over against a heap of natural resources that he uses. The independence of this character was always emphasized because the point of the picture was to show that he could manage without God. Typically it showed him as working a machine – driving a combine-harvester perhaps – with the crop that he cultivates and the landscape around it merely forming convenient extensions of its mechanism. Some of us, like the cosmologists whom I quoted earlier, have included other planets, as well as our present one, among these usable resources – handy spaces for our enterprise, or refuges that we can move to if things go badly with us here. (I think we need to take the myths of science-fiction as well as those of cosmologists, seriously because they shape people's imagination in a way that overflows into their lives). Instead of this dream, we need to learn somehow to see our species, and our culture, in their real situation as tiny dependent parts of an enormous whole, a vulnerable organic whole whose parts stand or fall together.

It's easy to say these things but the psychological journey involved – the shift that's still needed in our mental habits – is enormous. I will come back to this point about the unit of change later. But I think it's best first to say something about the other psychological obstacle, which is also large but in a way is simpler. This is the question of incentive. Can we live without that familiar prospect of future reward? Can we live in the present, no longer anticipating steady, continued improvements? Indeed – more alarmingly still – can we live without the prospect of always having at least our present level of comfort, convenience and medical care? Though we grumble about that current level it is, of course, something which, until the last century or so, virtually no human being ever enjoyed. Yet we have become so used to it that the idea of doing without it now is, to many of us, almost unthinkable.

Clearly the great strength of the Progress myth has lain in its offer of a safe and splendid future. This is something that we all hanker for when we have to do something difficult. Even in quite discouraging circumstances we usually do manage to find that hope somehow. But a widely-shared belief in a fixed salvation ahead is a great help to us. We need something we can trust. As William James pointed out, these confident hopes can often be self-validating. If two people have to leap across a gulf, one of whom believes he will be able to do it and the other believes he won't, they may both prove to be right.[3] I have suggested that the expectation of heaven used to served that purpose, and I think it's clear that, during the last two centuries, predictions of a mechanised heaven on earth have largely replaced it. Are we addicted to these long-term prospects? Can we find a way to live without them?

Steven Weinberg made a very interesting suggestion here. At the end of his book The First Three Minutes he raised questions about the purpose of life. He pointed out that current theories of physics predict for the universe 'a future extinction of endless cold or intolerable heat' and he concluded that this means the cosmos has no meaning. As he put it, 'the more the universe seems comprehensible, the more it also seems pointless'. This raises an intriguing question – Are we sure that the remote future is really so crucial? Supposing that, after the next major discovery, physicists changed their theories and told us that the universe may, after all, very well go on for ever, what would follow? Would that discovery prove that it does have a meaning after all, thus entirely changing our situation?

Something has surely gone wrong here about the notion of point or meaning, Weinberg is operating with the pay-off pattern whereby the point of anything is simply the reward that will follow it. This is appropriate when we are turning a handle to grind coffee, but not when we are playing a game or singing a song. The first part of the song is not a means to its last notes, nor is it a means to the drink that we may get after singing it or the fee that may follow. Fees and drinks are extras, not the central point of the activity. Nor is the whole process of the game merely a means to winning it, even though some people sometimes mistakenly treat it that way. Basically, we play or sing because we want to do so. These activities are ends in themselves This does not, of course, mean that they are isolated. The point or meaning or value of a song lies in its place in a larger whole – its connections with the rest of life. We value it because it enriches our wider sense of the pattern which connects everything.

This becomes clear even in the way in which Weinberg himself takes his gloomy pronouncement. He evidently doesn't mind much personally about the unfortunate mortality of the universe. He explains that he finds consolation for this cosmic futility in the work of astrophysics itself. He is content to know that, as he puts it,

[3]William James, essay on 'The Will To Believe' in volume with the same title, Dover, New York, 1956.

scientists 'build telescopes and satellites and accelerators, and sit at their desks for endless hours working out the meaning of the data they gather'. So it seems there is a meaning after all. Like the rest of us, Weinberg finds the point of his own life in his current occupations, not in an unimaginably distant future. But his words also imply that those data do have a point and a meaning, even if the universe as a whole does not. That point or meaning clearly doesn't lie in something that follows after them. It lies in the pattern of connections that makes sense of them.

There is a real paradox here that affects the current exclusive glorification of science. Can this occupation itself be so valuable if the subjects that it studies are themselves valueless? If the universe is not – as earlier scientists supposed – a glorious creation, testifying to the splendours of its creator, but a chance, meaningless muddle, why is it worth while to analyse it?

I stress this example of Weinberg's because I think the pay-off pattern that he uses has been over-emphasized in recent times in a way that is seriously misleading. The economic attention to means distracts us from ends – from all the things that really make life worth living right away. We are so focussed on certain future consequences – on the profits expected from particular actions – that we forget to ask crucial questions about what is happening now.

6 Faith, Values and Economism

We might ask here how people really thought about the prospect of Heaven (or Hell) in the past. Sometimes they ignored it. Sometimes they treated it as a practical issue about what they could get away with – perhaps by paying for almshouses or masses for their souls. But, besides this prudential angle, they plainly often thought of these prospects just as one aspect among many of their relation with God and their response to the ideals that went with it. Their notion of the future expressed the values that mattered most to them in the present. For many of them, the love of God was not so much an insurance policy as a general attitude to life, one which made a real difference to how they behaved. In this way people's expectations for the future are often an expression of what they most deeply believe in rather than a conclusion from factual evidence. And this influence of faith on conduct is not something obsolete which now gone away. For instance, economic policy in the last few decades has evidently been affected by many influential people's deep, heartfelt belief in the markets. That belief has produced a whole range of actions which less devout, more rational thinking would probably never even have considered.

These people have seen the market as an enclosing system that was taking them the right way, a ship to which they were rightly committed. It is quite close to the sense in which people believe in democracy, or the sense in which they used to believe in the British Empire. And until lately they certainly did not believe in this way in the planet, or in any of its ecosystems. These were simply resources which they took for granted, Thus, when tropical countries can make money from logging which destroys the rain-forest, international authorities have tended to see this as a quite proper form of development. If we then ask which organism is supposed to be developing here, it seems again to be that grand abstraction, the economy of the country. It certainly isn't the local ecosystem, nor the humans that form part of it.

The interesting thing here is how economic facts are treated as somehow more real than other kinds of fact. This priority is often expressed by phrases such as 'the bottom line', 'hard facts', 'at the end of the day', 'when you get right down to it' and indeed by a special use of the word realism itself. In this way economics, which is actually a rather abstract social science, is seen as a more reliable source of knowledge than enquiry about physical details such as the state of the soil or the health of the people who live there. This is what happens (as I suggest) when people believe in the economy in a sense which is much stronger than just thinking that it exists. No doubt they know that the rainforest exists as well. But the rainforest is less real to them than the balance of payments because it isn't a part of the guiding pattern by which they judge possible policies. They can't take it quite seriously. In predicting the future they are guided much more by that habitual pattern than by any more objective standards,

7 Myth and Reality

I have described these guiding patterns as Myths, which does not, of course, mean that they are lies. They are strong imaginative visions of a kind that we must have to shape our thought, to pull together its endless details into some necessary coherence. We need to use them, but they are always provisional. They have to be corrected by further insights, often arising from further facts.

The myth of Progress itself has been one of these visions. It was originally drawn from real facts about the rise of science and the successes of the Industrial Revolution. But as part of that general myth, there gradually grew also a subsidiary idea that the true measure of this development was money – that only economic arguments are truly realistic. This idea was, of course, important to Marx – which is why, in recent times, Marxists have so often turned into monetarists. But Marx himself included in economic facts the physical facts about the world which determine what can be traded. It is his more recent followers who have made the still more startling discovery that money itself is what really determines everything.

This, I am suggesting, is the dream from which the world's rulers are now, very gradually, beginning to be woken by that monotonous tap – tap – tap of environmental bad news. Plenty of them indeed are still locked in denial. But an increasing number have started to notice that climate change has turned out not to be just a glitch which scientists will shortly fix but a crucial, lasting fact. This change is deeply disturbing because it shows that this whole map is out of date. The habit of treating that very odd entity, money, as more solid than the things that we buy with it has had great attractions, but it clearly is no longer rational. The difficulty then is, how are people to think instead? What pattern can they now use? What vision should they now put their trust in?

Questions like these are often answered by experience. If the rain which is coming through the roof actually lands with a plop on the dreamer's shoulder this fact sharply alters his sense of priorities. In fact it alters his ontology, his whole sense of what is real. He suddenly loses interest in continuing his dream and responds instead to the immediate emergency. In the same way if we, or the world's rulers, were living on a small Pacific atoll, or in a country that is fast turning into desert, we probably wouldn't need to be told that the climate emergency is real. It is because we don't live there that we tend to react like the first-class passengers on an ocean liner who are visited by some ambassadors from steerage. The ambassadors report that the ship is sinking, but we tell them 'Not at our end', and go back to our cross-word puzzles. No doubt when this news is brought repeatedly we do gradually start to believe it, but we still can't see how to find a place for it in our background vision of life. We simply don't expect distant parts of the earth to form part of the business of our lives at all, still less distant parts of the earth's atmosphere. All this still sounds like somebody else's business.

8 The Need for Gaian Realism

If we are ever to get this right we shall probably need to make proper use of the concept of Gaia. Forty years ago, when James Lovelock first launched that concept, much of the learned world denounced it as unrealistic, fluffy and misty, in short, New Age, and not part of science. Since that time the concept's scientific merits have become clear and people now do see it that it's perfectly usable. But they still find a difficulty in actually using it.

The initial rejection flowed from a number of sources, but one powerful one was an affronted sense that this perspective was offensive to human dignity. The idea of seeing ourselves as just one tiny, dependent part of a vast organic body, rather than the sole agents present, the powerful owners of a great mass of resources, grated on the humanolatrous tendency which had been central to Enlightenment thinking. Very much the same thing happened to Darwin and indeed Lovelock's vision is a direct development of Darwin's. As the cartoons of the day show, public opposition to Darwin's ideas sprang much more from offence at the idea that humans were descended from apes than it did from any offence about God. The doctrine of primate descent threatened to remove humans from their position of dominance and put them right back inside the wider natural community which they both despised and feared. They are still resisting this sense of demotion. Many people, even people who reject religion, still want to claim a kind of extra-terrestrial status.

Thus this isolation of humans from the rest of life – this insistence on our total uniqueness which modern humanists feel is so crucial to our dignity – is linked to the general narrowing of our world-picture which I mentioned earlier. Because religious thinking had often proved tyrannous, we of the Enlightenment have, ever since the Renaissance, tried to assert a kind of human independence which was primarily aimed at providing freedom from God, but which also turned out to distance us from the natural world as well. We have felt that we are pure minds, alien to the material systems which we use and organize. Nor has our belief in evolution really shifted this stance because we have seen evolution itself as a pyramid from whose summit we can take off – not as a workplace in which we have a part to play,. Herbert Spencer's dream of evolution as a form of Progress – a race towards excellence, which we humans are winning – has had much more influence with us than Darwin's quieter notion of it as a radiating bush, producing life of every kind. Its past developments have seemed to be primarily a preparation for our successes.

Lovelock has startled us by a quite different vision. He has pointed to an activity of the earth in which we need to play our part – a drama where we have an active role. This drama has always surrounded us but we have managed to ignore it. It is interesting that Lovelock's insight into it emerged in the course of a particularly ambitious human enterprise – the investigation of Mars. Lovelock was working for NASA and was trying to find what would be the best indication of life on that planet. He reasoned that, rather than scooping up bits of Martian soil, it might help to find out about the planet's atmosphere. Earth, which does contain life, has an extraordinary atmosphere – a complex mix of many gases which are constantly interacting and being renewed. Yet, through all this activity, these interchanges remain so balanced that the conditions that make life possible have been maintained through three-and-a-half eons, ever since the first living things appeared. They have persisted even in the face of drastic changes, such as a great increase in the heat sent out by the sun. As Lovelock says, 'for this to have happened by chance is as unlikely as to survive unscathed a drive blindfold through rush-hour traffic'. What could possibly be keeping the planet in this life-friendly state? As he explains –

> Our results convinced us that the only feasible explanation of the Earth's highly improbable atmosphere was that it was being manipulated on a day-to-day basis from the surface, and that the manipulator was life itself. The significant decrease in entropy – or, as a chemist would put it, the persistent disequilibrium among the atmospheric gases – was on its own clear proof of life's activity.

That is, living things, as they breathe and transpire and excrete, apparently form part of a single tremendous mechanism, a living fountain which continually renews itself.

Somehow, then, the planet as a whole was doing this. Does this mean that it must itself in some sense be alive? Is this compatible with a proper definition of life? As Lovelock considered this possibility he looked at the literature and consulted with colleagues, but everywhere he found only vagueness and a general lack of interest about this concept and its possible limits -

> Take the concept of life. Everyone knows what it is but few if any can define it. It is not even listed in the [standard] Dictionary of Biology. ... If my scientific colleagues are unable even to agree upon a definition of life, their objections to Gaia can hardly be rigorously scientific. ... To a geophysiologist, a living organism is a bounded system open to a flux of

> matter and energy, which is able to keep its internal medium constant in composition and its physical state intact in a changing environment; it is able to keep it in homeostasis. . .. Gaia would be a living organism under the physicist's or the biochemist's definitions.[4]

He does not want to extend this thought anthropomorphically as some of his followers have done. As he says, 'When I talk of a living planet. I am not thinking in an animistic way of a planet with sentience. . . . I think of anything the earth may do, such as regulating the climate, as automatic, not through an act of will, and all of it within the strict bounds of science'[5] Yet he sees clearly that so dramatic a new conception must change our attitudes at a deep level. He writes, 'For me, Gaia is a religious as well as a scientific concept, and in both spheres it is manageable. . . .God and Gaia, theology and science, even physics and biology are not separate but a single way of thought'.[6]

These remarkable conclusions followed once he saw the significance of the contrast between Earth's atmosphere and that of other comparable planets. Mars and Venus have been found to have stable monolithic atmospheres, consisting mainly of carbon dioxide – conditions likely to make any sort of change impossible, and certainly to prevent the development of life. And at the time of his discovery the orthodox science of the day assumed that this was true on earth as well. The Earth's atmosphere was thought to have been originally produced by planetary outgassing and not altered afterwards except by abiological processes. In fact, atmosphere was viewed as just one part of the fixed, alien environment to which organisms were forced to adapt during evolution. The idea that organisms themselves might influence it – that they might help to provide the conditions for their own life – was quite alien to the science of the day. And any suspicion that these organisms, along with the atmosphere, formed part of one tremendous interactive process, so that the Earth could act as a whole, was, of course, still more alien to it.

Accordingly, when Gaian thinking first appeared scientists widely ignored it. Many non-scientists found it attractive, but this only made the scientific establishment more adverse. NASA lost interest in it once Gaian suggestions about Mars turned out to be unhelpful, so the new idea was left to make headway as it could against a scientific atmosphere where any reference to wholes in general tended to be looked at with suspicion. The success of atomistic thinking, both in physics and in biology, had accustomed many scientists to prefer explaining things atomistically, in terms of small particles, rather than by looking outwards to a wider context. And to think of the earth, with its atmosphere and all who inhabit it, as a whole required a quite different imaginative approach.

Since that time, many of Lovelock's detailed scientific suggestions have been investigated and many have proved convincing. Today, the notion that the living things of the earth do act together in this way is accepted as part of Earth Science – a new branch of learning which has been developed to bring together studies which

---

[4]Gaia; The Practical Science of Planetary Medicine, London, Gaia Books Ltd, 1991, p.29.

[5]The Ages Of Gaia; Oxford University Press 1988, p.

[6]GPSPM, pp. 206 and 212.

previously often knew little of each other's work. Yet in those quarters the name Gaia – which is that of the ancient Greek earth-goddess, mother of gods and men – is often avoided. Its symbolism may make perfectly good sense to the rest of us, but it causes alarm in the lab. Lovelock has therefore developed a less frightening form of imagery – the Medical Model. Earth, he suggests, is a sick planet – is, in fact, a patient who needs treatment – only unfortunately the experienced planetary physician who ought to attend her cannot be found. We shall therefore have to take on this case ourselves, even though we are inexperienced in such cases and indeed are ourselves a partial cause of the disease. We shall have to work out our diagnosis and treatment as best we can from what we discover, as physicians are often forced to do in new situations –

> We need this pragmatic approach now if we are to solve our planetary ills in time. We need planetary medicine. Its approach may be empirical, even at times unscientific, but it is all that we have. I am not proposing some kind of alternative science, the equivalent in medicine to acupuncture and homoeopathy .... My aim is merely to deflate the tumescence of big science and calm it down... If scientists are to recognise the value of empiricism in the troubled times to come, they must first acknowledge the extent of their ignorance about the earth... Modern medicine recognises the mind and body as part of a single system where the state of each can affect the health of the other. It may be true also in planetary medicine that our collective attitude towards the earth affects and is affected by the health of the planet.[7]

This picture, which shows the earth as a submissive patient lying in bed awaiting the doctor's opinion, certainly proved less threatening to scientists than the Goddess. Yet they still avoid the name, and often avoid the further thinking that is called for by any serious recognition of this entity as a whole.

Quite possibly, much of their difficulty here flows simply from the clash between modern specialization and the many-sided thinking that is needed to deal with so rich a concept. Today's scholars tend to acquire a sense that it is unprofessional to think about matters outside their own province, so they leave large topics alone. Thus John Ziman, wondering why scientists find Gaia so disconcerting, asks –

> Is this because it can't be squeezed into any of their established pigeon-holes? It mixes together concepts from the chemical, biological and physical sciences. ...
>
> I argue that this intrinsic pluralism is one of its glories and fascinations. Think historically. The planet Earth assembled, imbricated and remodelled itself by purely physico-chemical processes. For a billion years or so, everything that happened could be described in the language of gravitational forces, thermodynamic phases, chemical; compounds etc.
>
> Then, life emerged. Novel entities, with unprecedented properties – i.e. distinct organisms – appeared on the scene. To describe their phenomenology required a whole new conceptual vocabulary. Thus, the further history of Gaia had to be written, in part, in the language of biology. This had to include a great many absolutely basic terms such as organism, function, behaviour, metabolism [etc]... In due course, a million or so years ago, another conceptual fulguration occurred. The emergence of consciousness enabled hominids to engage in another completely unprecedented phenomenology.... with terms for social concepts... This, again, is strongly influencing the career of Gaia So yet another new language is required... So now we have to make sense of a world containing entities of

[7]GPSPM pp.14 and 71.

> these three different kinds, each governed by a different 'logic' and defined in a different conceptual language... And, because the successive events of their emergence were entirely unpredictable, as was what emerged at each stage, these phenomenologies, logics, languages and sciences are irreducibly distinct and cannot be unified into a single formal system. The pluralism of the sciences is not just a weakness of the human intellect; it is a product of the physico-biopsychic history of our Gaian abode.[8]

And this is indeed the challenge which this remarkable concept poses. We badly need scientists today who are prepared to deal with it.

[8]'The Challenging, Irreducible Pluralism of Gaia'. In Earthy Realism; The Meaning of Gaia, ed. Mary Midgley, Societas, Imprint Academic, Exeter 2007.

Equity: The Shortest Way to Global Sustainability

Vandana Shiva

My work on sustainability has been guided by some basic principles including the recognition that all members of the Earth community, human and non-human, have the right to sustenance, to food and water, to a safe and clean habitat, to security of ecological space. Resources vital to the sustenance must stay in the commons. The right to sustenance is a natural right because it is the right to live. These rights are not given by states or corporations, nor can they be extinguished by state or corporate action. No state or corporation has the right to erode or undermine these natural rights or enclose the commons that sustain life.

In India we have a very powerful concept of the *Earth family*: we are all members of the family of the Earth and therefore we must leave enough for others – that is the guide to action –, we must leave enough for others to maintain their share. We all have a duty to live in a manner that protects the Earth's ecological processes and the rights and welfare of all species and all people. No humans have the right to encroach on the ecological space of other species and other people or to treat them with cruelty and violence. And I think research must be subjected to that test of non-violence, too.

For me, sustainability is about not violating the rights of other human beings or other species. Long ago, just after the time I was involved in the Chipko Movement, I have done some research on mining impacts in the Doon valley, for the ministry of environment of India. And on the basis of that research the Supreme Court of India ruled that if commerce starts to undermine ecosystems and essential ecological processes that maintain life, then commerce must stop because the duty of the state is to protect, under article 21 of our Constitution – but I am sure there is similar articles in all constitutions. Article 21 requires the state to protect the life of all its citizens, and, if the conditions of life in terms of ecological conditions are being destroyed by some commercial activity, then that commercial activity must stop. This was the first legal ruling in India that on the basis of equity ruled on

V. Shiva (✉)
A-60 Hauz Khas, New Delhi, India
e-mail: vandana@vandanashiva.com

C.C. Jaeger et al. (eds.), *European Research on Sustainable Development*,
DOI 10.1007/978-3-642-19202-9_3,
Published by Springer-Verlag Berlin Heidelberg 2011

sustainability and linked the maintenance of ecosystems to the right of all humans to a secure ecological future.

That right to ecological space is for me the first principle of equity. The second principle of equity was reinforced at the Earth summit but it seems to have gotten forgotten since then. At the Earth summit we realized that the polluter must pay. That's equity: you made a mess, you clean it up, and you pay for its cleaning up.

Right now I am involved in a project on climate change in the Himalaya. The Himalaya is the third pole, people forget about it. The polar bear has become much more of a visual in terms of melting snows than all the Himalayan communities that suffer as glaciers melt, or all the communities in the drainage systems of the mighty Himalayan rivers. Nearly half of humanity would be hurt as glaciers melt. The snows are melting and everywhere more and more, I hear the conversation about how 'this weather is not usual', but that is what climate chaos means, nothing is usual. It's all unusual. And in the arid high Himalayan state of Ladakh, it is a desert, it never gets rain, it gets snow and when the snow melts it gets water, but now they started to get rain. And I have been in villages where in one downfall you can have a washing out of the entire village. Within half an hour, your life can be absolutely extinguished. And here are these innocent beautiful people of Ladakh punished by the actions of those who put all those contaminants and pollutants and greenhouse gases in the air. And no system of justice and equity to link that. We are still not linking it, because, amazingly, at the level of climate change, we have managed to evolve a system of the polluter gets paid through allocation of rights to the atmosphere and trade emissions. And I think we need to clarify how this is preventing equity in solutions to climate change and climate impact, especially with regard to vulnerable communities.

The third element of equity of course is the one that dominates international negotiations, so I will not spend too much time on that: it is about sharing of resources in terms of finances and technology. I could take any treaty and fill these two words as the expression of North–south equity.

There is a fourth aspect of equity that does not get noticed too much, it is the aspect of equity that became my learning to move out from of my profession as a physicist to doing the kind of research I do, in service of the Earth, in service of community, and that is the dimension of equal respect. It is about recognizing that every one is worthy of respect. And respect includes the recognition that no human being was put on this Earth without a brain. The brain is not a monopoly of those who get paid a salary to do research. Everyone has knowledge. Everyone has a mind, and we need to find ways to bring that knowledge to bear, because very often that is the knowledge that counts. My work on women and ecology has grown out of the fact that because women are the most vulnerable, because women carry the highest environmental burden – when wells go dry, they are the ones who walk extra miles – they are like the canaries in the coal mine and it will be good to talk to the canaries before a disaster in the coal mine.

There are, of course, major threats to equity, because many of these principles are systematically being violated. The first principle of the right to equal ecological space is violated by identifying prosperity and progress with higher levels of consuming the resources of the planet in an unequal way. I do not think it can be progress if 80% of things are used once to be then thrown away. I cannot see how that is progress. My mother still knew how to use and maintain clothes so that they can go on and on and on, and when they tear you can make something else out of them, and when that tears, you can turn it into a cleaning rag. For me, that is a sustainable economy, where everything can go on and on and on and ultimately get recycled and become the beginning of another cycle. By and large the throw-away economy is also an economy of pollution and waste. And I have watched in the last 10 years my country drown in the waste of the plastic and the aluminium that suddenly has become the symbol of living better. Consumerism ends up taking the resources of people away.

Moreover, 80% of the resources are used by 20% of the people. That means that 80% of the people go without adequate resources. And by going without adequate resources, they are unfortunately, then, suffering two kinds of consequences. One, being denied the right to live. You add up the numbers of people threatened from lack of food – one billion, another billion without water. But there is another connection, a connection that is often not made. That very often resource shrinkage, ecological depletion leads to resource conflicts, but these conflicts are not named as ecological conflicts. They are named as religious conflicts, they are named as ethnic conflicts. What is happening in Sudan is very much an issue of shrinking and disappearing resources. And you could do that in every part of the world, my early work on the green revolution in Punjab, on why Punjab was erupting like a volcano of terrorism in the eighties. And it came down to the fact that the resources of agriculture had actually shrunk with the application of non-sustainable technologies. The water had disappeared, the soil fertility had disappeared, the biodiversity had disappeared, work on farms had disappeared and collectively, this lead to discontent. It just so happens that the people of that area are Sikhs. And I think, if we were to do similar exercises for why did the Tamil Tigers arise in Sri Lanka as they did, part of it will be related to ecological inequities, some of it will be related to economic inequity and to social inequity. But it then inflates to this diversionary look at purely religious and cultural identity and then we get this definition of clash of cultures when really, what we are talking about is conflicts over very, very limited resources on this planet.

And as climate change and the end of oil, end of cheap oil at least, starts to make these issues even more important, we get the poor suffering three times over. The poor are victims, because first, they are displaced from work when a capital-intensive, fossil-fuel intensive system comes in, and the people's economies are destroyed. When giant industrial farms take over, farmers are displaced. When a highly integrated textile industry comes in, weavers work is displaced. At every level, there is displacement of work. Then they bear the disproportionate burden of the cost of climate chaos, through extreme droughts, floods and cyclones.

And as solutions are found, from the eyes of the privileged, to climate chaos, then the poor suffer one more time over, because solutions like industrial fuels divert their land and their food to the rich. And we have seen that with the case of biofuels, as food became the fuel for driving cars, and we have seen it now in the case of the land grab that even the economists deplore, this issue of countries grabbing land in the south, some for food, but primarily for biofuel. And this land grab is in millions of acres. The heads of states, the leaders of governments who hand over this land never take the permission of the people who own the land. 'Can we give this to a Korean firm in Madagascar?' 'Can we give this to a UK biofuel firm, appropriate it in a violent way and then hand it over?' Japan is proposing a G8 meeting on addressing this issue of the land grab. We have just had the recent elections in India, and since I know most of the areas where this land conflict is a big issue, I am very happy to say that every politician who supported the movement of peasants and tribes in defence of their land, that politician won, no matter what the party. So it is literally a vote for ecological equity. It is a victory for land sovereignty.

Without equity, sustainability will definitely go wrong. It will first go wrong because you do not have the social base for conservation. These resources are commons, it is only when you are organised as a community that you can protect the forest, that is what I learned in Chipko. It is only when you are organised as a community, that you can do water harvesting so that the river water will be shared equitably. You can exploit resources under unequal regimes. You cannot conserve resources under unequal regimes. It is an asymmetric relationship. You can have increasingly high levels of extraction of resources and their conversion to commodities. And much of what has been called efficiency, much of what has been called productivity, was related to efficiency and productivity of resource exploitation. I think it is time for us to start feeding in efficiency and productivity in terms of systems of resource conservation. It will take a challenge because linear thinking would have to be turned into circular thinking, and forgetting about the costs of resources and forgetting the cost of pollution will have to now be internalized – a much more difficult job than a linear flow of resource extraction and pollution dumping.

There is a second reason why equity becomes very important to sustainability: because commoditisation creates what I have called the monocultures of the mind. Monocultures of the mind focus on a singular entity. In a forest with 5,000 species they look at the one timber species that is valuable and then maximise its production. In farming, instead of the 8,500 crops we have grown, the focus is on soya and corn and canola and cotton as if humanity did not need anything else. And then you get larger and larger acreages. There is a fascinating ad Syngenta: 'How do we feed the growing world population – farm new land or get more from existing land?' And interestingly every time they talk about avoiding farming new land, there is more farming of new land. The Amazon becomes farmland, the Indonesian rainforest becomes farmland and what is called intensive agriculture ends up being extensive agriculture because biodiversity was forgotten.

In the work I have done over the last 25 years in India, both at the level of practice as well as at the level of research, we have found that the more biodiverse the system is, the smaller the farms are, which means more equity in land distribution, more equity also in seed entitlement – farmers having their seeds rather than intellectual property turning seed into a monopoly of five companies. The more biodiversity, the more seed sovereignty, the more land distribution and land sovereignty, the higher food production and the higher the incomes.

Finally, there is the coal issue. In our work on organic farming, we are doing research on how much carbon capture can happen in the soil to increase the fertility of soils, to mitigate climate change, to help small, vulnerable communities to have better adaptation capacities when a severe drought hits them or a severe cyclone hits them. Four years ago we had a drought in central India, rain fell from 900 to 400 mm. Only the farms that we had trained to go organic could get a crop over the 4 years with just the soil moisture because of the organic content. Eight to ten years is when the CCS (Carbon Capture and Storage) systems will deliver. Organic farming turning the soil into the real sink can deliver yesterday, has delivered yesterday and the day before. And I think we should not forgo these proven options, because they are the ones that work for the poor. CCS will work for the few hand full of power utilities, organic farming will work for half of humanity in production and for all of humanity in the entitlement to good food.

We are at a very critical moment, we are amidst a democratic transition from an oil-centred, capital intensive system of production, to a soil-centred, people-centred system of production. The poor, the weak, the excluded, the marginalised are threatened today. In the short term, we can continue to extend the profits, the consumerism of the privileged, by further dispossession of the poor. But tomorrow, even the rich and the powerful will not be immune from Gaia's revenge and the revenge of the billions of dispossessed. We will either have justice, sustainability and peace together, or we will descend into ecological catastrophe, social chaos and conflict. As I said in my most recent book 'Soil not Oil': 'Soil not Oil offers a framework for converting the ecological catastrophe and human brutalisation we face into an opportunity to reclaim our humanity and our future and build it together'.

The Sustainability Concept: Can We Stand Between Catastrophism and Denial?

Nicole Dewandre

A talk inspired by Hannah Arendt[1]

1 Out of Daily Practice, an Exercise of Thought

Having been the head of a unit entitled "sustainable development"[2] in the Directorate general for research at the European Commission, it is obvious that I lived with this word "sustainability" most of the day. Let's be clear: I will not define sustainability. We are well aware that sustainability is an ambiguous concept. In the EU, we have – as you know – the renewed EU Sustainable Development Strategy (renewed EU SDS).[3] This allows us to cut short the definitional issue and adopt a pragmatist point of view: from an institutional perspective, sustainable development is . . .what is in the sustainable development strategy, i.e. tackling the seven key challenges[4] in an integrated manner!

And the mandate of the "sustainable development" unit in DG Research has been to optimise the articulation between this renewed EU SDS and EU-funded research, i.e. the so-called 7th Framework Programme (FP7). By the way, let me

[1]The opinions expressed are those of the author only and should not be considered as representative of the European Commission's official position.

[2]This unit has existed between October 2006 and December 2010. It has been dismantled in the reorganisation that entered into force on January, 1 2011.

[3]http://ec.europa.eu/sustainable/docs/renewed_eu_sds_en.pdf.

[4]The seven key challenges are: 1. climate change and clean energy, 2. sustainable transport, 3. sustainable consumption and production, 4. conservation and management of natural resources, 5. public health, 6. social inclusion, demography and migration, 7. global poverty and sustainable development challenges.

N. Dewandre (✉)
Information Society and Media Directorate-General, European Commission, Brussels, Belgium
e-mail: nicole.dewandre@ec.europa.eu

C.C. Jaeger et al. (eds.), *European Research on Sustainable Development*,
DOI 10.1007/978-3-642-19202-9_4, © European Union, 2011
Published by Springer-Verlag Berlin Heidelberg 2011

inform you that we have set up a monitoring tool that allows cross-referencing between FP7-funded cooperative research, on the one hand, and all the operational objectives of the renewed EU SDS, on the other hand. This web-based tool has been operational since April 2010[5] and allows the identification of the bulk of EU-funded cooperative research contributing to each operational objective. With this system, we can measure that 66% of the FP7 cooperative projects flowing from the first four waves of calls for proposals (2007–2010) contribute positively to one or more objectives of the renewed EU SDS. The monitoring system allows us also to focus on specific grand challenges, such as the low-CO2 society. Almost half (40%) of the allocated budget for cooperative research after the first four years of FP7 implementation (2007–2010), or 4.27 billion €, is allocated to research conducive to a low CO2 society. The system can also zoom in on an operational objective, and consequently identify the group of FP7 projects contributing to that objective. For example, 474 million EUR from the EU budget (again over the period of 2007–2010) have been allocated to 82 projects contributing to raising the share of renewables. The biggest part of this research comes from the budget earmarked for energy research, but other thematic areas contribute also to this effort: agricultural research, materials research, transport research and space research. In some sense, this web-based monitoring tool can be seen as an invisible mind...

This being said, what follows in this contribution are some personal thoughts about "the sustainability concept", as was nicely offered to me by the editors of this book. Of course, these thoughts are grounded in and nurtured by my professional experience, but as any exercise of thought, it cannot represent an institutional point of view, as thinking is necessarily an individual idiosyncratic activity. Let me be clear and explicit about the fact that I am greatly inspired by Hannah Arendt. There is no need to know her work to understand what follows, but those acquainted with her philosophy will of course recognise the source of inspiration.

2 Moving Beyond Catastrophism and Denial

The public space of our times, at least as I feel it here in Brussels, is full of a "collective anxiety" about what the future holds for us collectively at the most general level – humankind – and also at more specific levels: "western civilisation", nationalities, regional identities, our families, ourselves as individuals. It is as if all our challenges have to be perceived through fear and outcome-to-be-avoided, rather than in terms of positive outcomes; as if we had lost collective confidence. I see the resonance of catastrophism in the political arena as a corollary of this situation. The defenders of catastrophism consider that alerting people to visions-to-be-avoided is

[5]www.fp7-4-sd.eu.

the best – if not the only – way to raise awareness and induce the desired changes in behaviours.

If this were true, i.e. if catastrophism had this efficiency, I'd have no problem with it: I am a pragmatist! But the problem is that catastrophism has deep and perverse side-effects. Indeed, creating a vision, . . .in order to avoid it, works only in very specific situations that we all experienced as children, or later as parents. It goes like this: "If you cross the street without looking carefully first, you will have an accident", or "if you go too close to the cliff, you will fall". But in the political arena, creating visions-to-be-avoided does not work. Collectively, we go to the visions that we create for ourselves. Visions are performative. Societies or civilisations cannot behave like a super-"stunt man" in a James Bond movie: avoiding the worst outcome at the last minute! This is science-fiction, very useful as such but must remain as such. Catastrophism, despair, drama are part of society. It has always been part of the arts' register. Apocalypse is at the core of artistic inspiration. It's a visual spectacle, it has strong impacts, and it provides a fantasy. People are attracted by death and drama. So, one can say it is legitimate to use catastrophism in arts. However in the politics' register, there are – at least – two risks linked to the use of catastrophism: (1) credibility of public action and (2) totalitarianism. Regarding the latter (totalitarianism), catastrophism aims to mobilise around fear. This creates the perfect conditions for dictatorship. Regarding the former (credibility), the abuse of the argument ("last chance", "now or never", etc.,) leads to the situation where institutions have to face the day after and keep on working towards the same goal! Cancun after Copenhagen. . . Or is it really true that we'd better turn around and forget about climate change because it's too late anyway? Hence, my deep conviction is that catastrophism has two big drawbacks: by inspiring fear, it leads only to despair and – maybe – to totalitarianism; by alarming people, it creates powerlessness and puts the credibility of public action at risk.

Yes, but. . ."What else?" can we then ask ourselves.

The alternative against which catastrophism is mobilised can be called the denial attitude. "Let's not worry", "let's continue and do business-as-usual", "business will overcome these problems by change and innovation". But here we face another paradox related to change and innovation. There is a way to speak about change and innovation which is particularly . . .steady and non-innovative, a sort of "blind rhetoric". This discourse on innovation is also often loaded – let's be frank – with some "victimisation" from industry: there is too much regulation or administrative burden, society does not like innovation, etc., . . . This blind discourse on the need for change and innovation is – when looked at more closely – just the other side of the "catastrophism" medal. It does not provide for a vision. It expresses only the fact that time passes by, whether we like it or not!

This confrontation between catastrophism and denial is sterile. It does not allow us to grasp what has disappeared and seems out of reach: the sense of the future, the purpose of our collective endeavour.

3 Articulating Limits and Openness: Endorsing Contingency in Politics

This confrontation between catastrophism and denial has only one advantage: it defines "sustainability" by its boundaries. By this, I mean that sustainability is what lies between catastrophism and denial. The challenge is to fill this concept with a positive meaning, reclaiming a collective sense of purpose. How can we shape the "sustainability concept"? Let's see what can be taken from catastrophism and denial!

From the catastrophist standpoint, we can keep the notion of limits. It seems that men have always experienced problems with thinking about limits and coping with them. Limits seem to automatically trigger the will to transcend them: seeking eternity, invading neighbours, exploring unchartered territories. Curiosity has always been linked to going beyond limits. And the need to transcend limits has not only been recognized, but highly praised in Western civilization. Limits were to be pushed back, defeated. This was the essence of the faith in progress, on which this Western civilisation has evolved during the two last centuries. Hannah Arendt drew our attention to the fact that, since the Shoah and Hiroshima, this blind faith in progress based on transcending limits has disappeared. Whether we like it or not, this faith in progress is now a "vestige" of our political culture, but not an active principle anymore. We know we have to keep this compulsive reaction of transcending limits at some distance. We have to learn to deal with limits without necessarily wishing to transcend them. It's all about recovering wisdom and prudence as virtues. To succeed in doing this, it is my intuition that we need to hear more women's voices.

From the denial standpoint, we can keep the notion of openness. A wrong way to learn to deal with limits is to see limits everywhere!!! Or else to think that they are hidden and we must find them out. The future is intrinsically open – philosophers will say "contingent", as opposed to "necessary". In a sense, the catastrophist viewpoint can be seen as the (ill-defined) transposition into time of the limitation of space, or the (still ill-defined) transposition in the world of the limits of our individual lives. What defines the world is precisely that it will survive each of us individually, says Hannah Arendt. It is essential for political action to rest on this worldview of openness and durability of the world.

4 Is the Future to Be Grasped with Knowledge?

This double recognition of limits and openness leads to the question of the role of knowledge when it comes to predicting the future, and corollarily to the role of knowledge in policy-making. Ultimately, it is about thinking of the triangle made of knowledge, power and action. we often hear that we need to deal with uncertainty. More than uncertain, the future is unknowable. Of course, we know (some) things about the future. But what we *know* about the future is not what makes history. What we know conditions and shapes our way to apprehend the future, but, fundamentally, future – in the political sense – is not and cannot be an object of knowledge.

The future in the political sense of the term is driven by a collective sense of purpose. For the moment, it can be said that the collective sense of purpose is somewhat lost, and catastrophism and denial are only default solutions to capture the quest for sustainability. Sustainability is – to some extent – synonymous to "reclaiming a collective sense of purpose"! Harnessing knowledge to sustainability means *in fine* reconnecting knowledge with confidence in human action and in the future. As paradoxical as it may seem, reclaiming the sense of the future passes through asking the question of meaning in the present time. Indeed, we know that the path to long-term horizons is nothing more than a succession of short term ones. Trying to escape the present times is just stepping out of the world. We are not only "*earth-bound*" creatures, but also "*present-bound*" ones. Asking the future to be the reference point from which best options can be identified is, in some way, trying to look at the world from a God's perspective. Let's always keep in mind the following question: "how do my actions contribute to shaping a collective sense of purpose here and now?" This is a question underlying most of the contributions of this book and its brother "A new quality of life."

5 Integration and Diversity in the Light of *Amor Mundi*

Another way to look at what it takes to give shape to sustainable development is to consider two of the most common concerns expressed in circles dealing with sustainability issues: the need for more integration, on the one hand, and the need to protect biodiversity, on the other hand. First, let's take note that the lack of integration is mentioned in any classification that is put forward: the famous so-called three pillars, the scientific disciplines, the gap between science and policy, etc., … Behind this generic diagnostic of a lack of integration, which resembles furiously a truism, lies necessarily the preconceived solution, i.e. the integration of what is deemed to be non-integrated would lead to sustainable development. But as William James already pointed out more than 100 years ago, there is no definitive "whole" that can be grasped. All views, thoughts and postures are necessarily partial and temporary. The claim for integration as "the" solution is blind and deaf to this basic lesson. However, in line with pragmatism, we recognise that there is a way to make sense of this claim for integration, with the following considerations in mind:

1. There can't be knowledge or collective action without the creation of categories and typologies.
2. Specific categories and typologies are contingent, i.e. in a given context, they are what they are, but they could have been otherwise. This does not mean that any

typology would necessarily make sense, but only that there are several equally sound candidates.

3. For any given categorisation and typologies, there is not only room but need for cross-cutting approaches. The need for cross-cutting approaches should not be seen as a failure or a weakness of a given typology.

With this in mind, we can redescribe the claim for integration as a claim for *constructive* interactions with the other constituencies, be they scientists from other disciplines, stakeholders voicing other concerns, colleagues from other services, etc.,. . . This redescribed claim for integration applies then first to the bearer of the claim and leads to a relaxed approach to all the forms of otherness, as defined by the multiple factors of identity of the locutor. Instead of instrumentalising interactions to reinforce identity and building homogenous collective entities, interactions should be considered for serving the concrete situation which calls for the given interaction. In other words, identity concerns should be *neutralised* as much as possible in the game of interactions.

Regarding the need to protect biodiversity, let's listen to Elinor Ostrom when she points to the fact that it is not only biodiversity that needs to be fostered, but all forms of diversity, and let's admit that it would not make sense to seek to protect biodiversity with a mental frame where all other forms of diversity are perceived as a lack of coherence, a weakness or a failure. In which terms can we redescribe this call for protecting diversity? Seeking diversity means welcoming differentiation and alterity. This mitigates – or sets some limits – to some general principles which are active in the public sphere or in science: the principle of objectivity and equal treatment, the search for general laws, the paramount logic of rationalistic optimalisation. Protecting diversity entails coping with differences, *as they emerge*, in a positive way instead of fighting them or trying to reduce them under general characteristics. Differentiation (coping with differences as they emerge) is to be radically distinguished from discrimination (making differences and building walls). And this is also where we can hear Hannah Arendt whispering into our ears that the only guarantee of the durability of the world lies in the continuous stream of new births of men and women, and our capacity to decide if we love this world enough to let them undertake something that we did not predict, and even not foresee.

To conclude, going back to the original question: "*can we stand between catastrophism and denial*?", my answer is . . ."That's the only place to be for the living human beings that we are".

Social Sustainability: Exploring the Linkages Between Research, Policy and Practice

Andrea Colantonio

1 Introduction

In recent years the social dimension (or 'social sustainability') has gained increased recognition as a fundamental component of sustainable development, becoming increasingly entwined with the delivery of sustainable communities discourse and the urban sustainability discourse. Environmental and economic issues dominated the sustainable development debate at its beginning whilst it is only in the late 1990s that social issues were taken into account within the sustainability agenda. Although its growing recognition has spurred an emerging body of literature on social sustainability, our understanding of this concept is still fuzzy and limited by theoretical and methodological constraints stemming from its context and disciplinary-dependent definitions and measurements. As Sachs (1999) puts it, at a fundamental level, it is still unclear whether the concept of social sustainability means the social preconditions for sustainable development or the need to sustain specific structures and customs in communities and societies.

Thus, the aim of this chapter is threefold. Firstly, it provides an overview of the concept of social sustainability and explores its evolutionary meaning, highlighting the shift from the analysis of traditional 'hard' social policy areas towards emerging 'softer' research and policy-making themes. It is important to clarify that this chapter does not seek to provide operational definitions of, or normative prescriptions for, social sustainability. Rather, it debates alternative readings of social sustainability in the light of past, present and possible future interpretations of this concept. The second main objective is to examine to what extent and how social sustainability has translated into policy, despite the ongoing debate regarding the level of integration of assessment techniques, themes and metrics. Lastly, the chapter endeavours to suggest possible future trends in social sustainability

A. Colantonio (✉)
Research Coordinator, London School of Economics and Political Science, United Kingdom, Houghton Street, London WC2A 2AE, United Kingdom
e-mail: andrea.colantonio@gmail.com

C.C. Jaeger et al. (eds.), *European Research on Sustainable Development*,
DOI 10.1007/978-3-642-19202-9_5, © European Union, 2011
Published by Springer-Verlag Berlin Heidelberg 2011

research and ways in which the linkages between research, policy and practice should be investigated.

The chapter is divided in four main parts. It begins with an overview of the main interpretations of social sustainability that illustrates how different worldviews amongst social scientists have thus far prevented an unequivocal and widespread acceptance of the themes at the heart of this notion. The second part examines how social sustainability theoretical research approaches have translated into policy geared towards the promotion of social capital, citizens' participation, capacity building and, more recently, city liveability strategies. The third part illustrates how impact assessment is evolving into sustainability assessment (SA), and new appraisal methods and metrics are emerging in the sustainability literature. In this context, the analysis highlights the main differences between 'traditional' and 'sustainability' indicators, suggesting a set of characteristics for the latter. The chapter concludes with an examination of possible future directions within the social sustainability debate and the challenges that will have to be overcome to assess the progress toward sustainability.

2 Social Sustainability

There is general agreement that the different dimensions of sustainable development (e.g., social, economic, environmental and institutional) have not been equally prioritised by policy makers within the sustainability discourse (Drakakis Smith 1995). This is mainly because sustainable development was born out of the synergy between the emerging environmental movement of the 1960s and the 'basic need' advocates of the 1970s, but also because assessing the intangible nature of social aspects of development presents measurement quandaries, which will be discussed later. As a result, there is limited literature that focuses on social sustainability to the extent that a comprehensive study of this concept is still missing. Indeed, Littig and Grießler (2005) argue that approaches to the social sustainability concept have not been grounded on theory but rather on a practical understanding of plausibility and current political agendas. In addition, a recent study by the OECD (2001) points out that social sustainability is currently dealt with in connection with the social implications of environmental politics rather than as an equally constitutive component of sustainable development.

These fragmented approaches to social sustainability are also criticised by Metzner (2000) who contends that social sciences and social policy research have developed a plethora of social objective strategies and measurement instruments, but with little regard for the sustainability perspective. Thus, while there exists abundant social research studies and policy documents, these have rarely been integrated into the sustainability framework. Even when cross-discipline approaches have been attempted, covering for example the environmental and the social dimensions of sustainable development within the 'ecological footprint' concept (Wackernagel and Rees 1996), it can be argued that such endeavours have only been partially framed within an integrated approach to sustainability.

As a result, the concept of social sustainability has been under-theorised or often oversimplified in existing theoretical constructs and there have been very few attempts to define social sustainability as an independent dimension of sustainable development. For these reasons, it can be argued that the relationships between the different dimensions of sustainable development, or indeed between 'sustainabilities', are still very much unclear. For example, Assefa and Frostell (2007) contend that social sustainability is the finality of development whilst economic and environmental sustainabilities are both the goals of sustainable development and instruments to its achievement. Similarly, Hardoy et al. (1992) dispute interpretations according to which social sustainability is defined purely as the social conditions necessary to support environmental sustainability. Furthermore, no consensus seems to exist on what criteria and perspectives should be adopted in defining social sustainability. Each author or policy maker derives their own definition according to discipline-specific criteria or study perspective, making a generalised definition difficult to achieve. Nonetheless, several definitions are reported in Table 1, which provides an overview of the plethora of social sustainability interpretations.

In Table 1, it can be seen how in Sachs' views (1999) socio-economic development is an open ended historical process, which partially depends on human imagination, projects and decisions subject to the constraints of the natural environment and the burden of the living past. Thus, social sustainability can be

Table 1 Examples of definitions of social sustainability

A strong definition of social sustainability must rest on the basic values of equity and democracy, the latter meant as the effective appropriation of all human rights – political, civil, economic, social and cultural – by all people	Sachs (1999: 27)
...a quality of societies. It signifies the nature-society relationships, mediated by work, as well as relationships within the society. Social sustainability is given, if work within a society and the related institutional arrangements satisfy an extended set of human needs [and] are shaped in a way that nature and its reproductive capabilities are preserved over a long period of time and the normative claims of social justice, human dignity and participation are fulfilled.	Littig and Grießler (2005: 72)
[Sustainability] aims to determine the minimal social requirements for long-term development (sometimes called critical social capital) and to identify the challenges to the very functioning of society in the long run	Biart (2002:6)
Development (and/or growth) that is compatible with harmonious evolution of civil society, fostering an environment conducive to the compatible cohabitation of culturally and socially diverse groups while at the same time encouraging social integration, with improvements in the quality of life for all segments of the population	Polese and Stren (2000: 15–16)

interpreted as a socio-historical process rather than a state. In this perspective, the understanding of social sustainability cannot be reduced to a static zero-one situation where zero suggests an unsustainable situation and one indicates presence of sustainability.

From a strictly sociological standpoint Littig and Grießler (2005: 72) emphasise the importance of both 'work', which is a traditional anchor concept in the German sustainability discourse, and 'needs' as defined by the Bruntland Commission (1987). Similarly, Biart (2002: 6) highlights the importance of social requirements for the sustainable development of societies. Despite the confusion over the meaning of social capital, his approach emphasises the importance of 'time –frames' and 'social conditions' for the long term functioning of societal systems. However, in his analysis there is no reference to the physical environment, allowing for the traditional criticism that sociology has often suffered from a neglect of the physical and non-social realm (Omann and Spangenberg 2002).

A more comprehensive definition of social sustainability with a special focus on urban environments is provided by Polese and Stren (2000: 15–16). They emphasise the economic (development) and social (civil society, cultural diversity and social integration) dimensions of sustainability, highlighting the tensions and trade-offs between development and social disintegration intrinsic to the concept of sustainable development. However, they also acknowledge the importance of the physical environment (e.g., housing, urban design and public spaces) within the urban sustainability debate. Within the context of urban areas, other authors also maintain that social sustainability interpretations emphasising social equity and justice may assist cities in evolving to become 'good' places by facilitating a fairer distribution of resources and a long term vision (Ancell and Thompson-Fawcett 2008).

By contrast, from a narrower housing and built environment perspective, Chiu (2003) identifies three main approaches to the interpretation of social sustainability. The first interpretation equates social sustainability to environmental sustainability. As a result, the social sustainability of an activity depends upon specific social relations, customs, structure and value, representing the social limits and constraints of development. The second interpretation, which she labels 'environment-oriented', refers to the social preconditions required to achieve environmental sustainability. According to this interpretation, social structure, values and norms can be changed in order to carry out human activities within the physical limits of the planet. Lastly, the third 'people-oriented', interpretation refers to improving the well-being of people and the equitable distribution of resources whilst reducing social exclusions and destructive conflict. In her study of the social sustainability of housing, Chiu (2003) adopts the second and third approach to demonstrate how social preconditions, social relations, housing quality and equitable distribution of housing resources and assets are key components of sustainable housing development.

Other authors do not provide a general definition of social sustainability but suggest the main key themes at the basis of the operationalisation of this notion. A number of these key themes are listed in Table 2, which shows how basic needs and equity are consistently being held as fundamental pillars of social sustainability. These concepts are deemed necessary for the physiological and

Table 2 Key themes for the operationalisation of social sustainability

Feature	Author
• Livelihood • Equity • Capability to withstand external pressures • Safety nets	Chambers and Conway (1992)
• Inclusion • Equity • Poverty • Livelihood	DFID (1999)
• Equity • Democracy • Human rights • Social homogeneity • Equitable income distribution • Employment • Equitable access to resources and social services	Sachs (1999)
• Paid and voluntary work • Basic needs • Social security • Equal opportunities to participate in a democratic society • Enabling of social innovation	Hans-Böckler-Stiftung (2001)
• Social justice • Solidarity • Participation • Security	Thin et al. (2002), DIFD
• Education • Skills • Experience • Consumption • Income • Employment • Participation	Omann and Spangenberg (2002)
• Basic needs • Personal disability • Needs of future generations • Social capital • Equity • Cultural and community diversity • Empowerment and participation	Baines and Morgan (2004) and Sinner et al. (2004)
• Interactions in the community/social networks • Community participation • Pride and sense of place • Community stability • Security (crime)	Bramley et al. (2006)

social survival of human beings and communities as a whole. This is because, at a basic level there can be little doubt that shelter, food, clean water and employment are essential requirements for the sustainability of individuals and communities. Similarly, equity is considered a crucial component of social sustainability because

Table 3 Traditional and emerging social sustainability key themes

Traditional	Emerging
Basic needs, including housing and environmental health	Demographic change (aging, migration and mobility)
Education and skills	Social mixing and cohesion
Employment	Identity, sense of place and culture
Equity	Empowerment, participation and access
Human rights and gender	Health and safety
Poverty	Social capital
Social justice	Well being, happiness and quality of life

of the increasing evidence that societies with lower levels of disparity have longer life expectancies, less homicides and crime, stronger patterns of civic engagement and more robust economic vitality (GVRD 2004).

The chronological review of these themes suggests that traditional themes, such as equity, poverty reduction and livelihood, are increasingly been complemented or replaced by more intangible and less measurable concepts such as identity, sense of place and the benefits of 'social networks'. Table 3 illustrates this broad shift from 'hard' themes towards 'softer' concepts within the sustainability discourse, although it is worth clarifying that even traditional 'hard' themes such as ageing and migration are increasingly being approached from a more qualitative perspective. For example, in recent years the study of migration is not only limited to the analysis statistical figures but also entails the examination of the qualitative nature of migration and the level of integration of migrants in their recipient countries.

Despite these disagreements, for the purpose of this chapter, it can be argued that social sustainability concerns how individuals, communities and societies live with each other and set out to achieve the objectives of development models, which they have chosen for themselves taking also into account the physical boundaries of their places and planet earth as a whole. At a more operational level, social sustainability stems from actions in key thematic areas encompassing the social realm of individuals and societies, ranging from capacity building and skills development to environmental and spatial inequalities (see Colantonio 2007, for a complete list). In this sense, social sustainability blends traditional social policy areas and principles such as equity and health, with issues concerning participation, needs, social capital, the economy, the environment, and more recently, with the notions of happiness, well being and quality of life. The different role played by theories, principles, objectives, targets and themes in the purse of social sustainability will be reviewed in the remainder of this chapter.

3 The Linkages Between Research and Policy

Several theoretical and methodological approaches to the study of social sustainability can be identified. These include, for example:

- *Capital stock*, e.g., social capital, environmental capital and ecological footprints etc. (Coleman 1988; Putnam 1993; Wackernagel and Rees 1996).
- *Equity and Human Rights*, e.g., poverty studies and unequal development (Sen 1985, 1992; Sachs 2001);
- *Institutional Theory and Governance*, e.g., participation and stakeholder analysis (Chambers 1992 ; Healey 1992);
- *Business and Corporate studies*, e.g., Triple Bottom Line, Corporate Social Responsibility etc. (Elkington 1994);
- *Behavioural and Welfare Economics*, e.g., capabilities approach, well-being, health and happiness perspectives (Sen 1993; Nussbaum and Glover 1995; Layard 2005);
- *Transition Theory*, e.g., institutional theory and system analysis (Rotmans et al. 2001; Loorbach and Rotmans 2006).

Throughout these perspectives there has been an examination of different aspects of the 'social' dimension of development, but the exact positioning of this varies depending on the perspective adopted. For example,

- The equity and human right approach emphasises the inter and intra-generational aspect of the benefits of development;
- Advocates of the capital stock perspective focus on the importance of human relationships (social capital) and its impact on development or the physical carrying capacity of planet earth (environmental capital) and how these capitals are shared amongst individuals and societies across the globe;
- Other authors use institutional theory to highlight the importance of participation in the governance mechanisms underpinning development;
- The business approach calls for a more ethical and pro-active role of the private and corporate sectors in improving the social qualities of communities and places where they operate;
- Behavioural and welfare economy scholars have recently pointed out that both the pre-conditions and the finality of development should be to increase people's happiness and quality of life, whilst;
- Transition scholars have recently focused on the elements required to foster a systemic societal shift from the current unsustainable development model toward a more sustainable one.

It can be argued that these approaches have been incorporated into sustainable development policies promoted at both national and international level to different extents. Figure 1 shows a few examples of how social sustainability research approaches have led to the promotion of specific policies. These include for example the promotion of social capital by the World Bank; capacity building programmes promoted by the UN, and urban development policies geared toward the enhancement of quality of life and happiness in several EU cities. In addition, more recently, the Lisbon European Council held in 2000 also launched for the first time the idea of the social dimension as an integral part of sustainable development, paving the way for the Bristol Accord and the EU sustainable communities agenda in 2005.

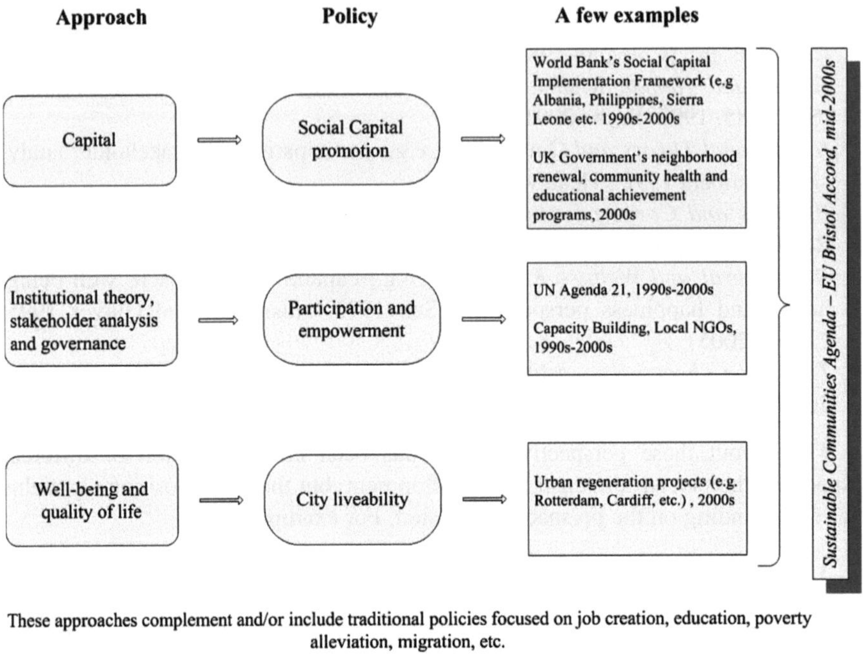

Fig. 1 Research approaches and policy linkages

However, the mechanisms through which new theories and approaches provide stimulus for new policies are still unclear. Further, there is disagreement on how long it takes for new research ideas to be incorporated into policy prescriptions. This can range from a few years to generations of policy makers. Indeed, the linkages between research and policy are influenced by several elements, including

1. *Level of abstraction of the theory*, which may have little application
2. *Feasibility and implementation costs*, including for example the cost of maintaining sets of local indicators to monitor the effectiveness of policies
3. *Complexity and sophistication*, which may preclude access to and participation of several stakeholders due for example to the existence of technical jargon etc.
4. *The nature of the dialogue and communication channels existing between researchers and policy-makers*

Furthermore, at a more conceptual level, it can be argued that another element influencing the linkages between research and policy is the shift from 'hard' themes towards 'softer' concepts within the social sustainability discourse, which has spurred a wider debate on the role that governments and policy-makers should play in delivering 'soft' objectives. For example, with regard to happiness, Ormerod and Johns (2007) question the ability of governments to embark upon happiness-oriented policies whilst they are still struggling to deliver on existing commitments. By contrast, Layard (2007) notes that governments have been interested in happiness at least since the Enlightenment, but only recently they have begun to measure the concept and explain it systematically. Thus, understanding the conditions conducive to human happiness in all their complexity should be the central concern of social science.

Due to the speculative nature of social sciences, and the emerging mix of hard and soft themes in the social policy debate, it is may prove difficult to scientifically understand the inter-relationships between these themes and to identify optimal social targets and objectives to be pursued in order to deliver socially sustainable places. Indeed, the multiple combinations of hard and soft themes, and the disagreement over their meanings, hinder the scientific identification of what is socially sustainable and what it is not.

More dangerously, ill-conceived assumptions and theories concerning the elements conducive to social sustainability can potentially lead to the implementation of inadequate social policies. A classic example of this peril is represented by the assumption that higher income automatically fosters more socially sustainable communities, for example, by reducing crime or boosting personal and communal satisfaction. However, there is evidence in EU cities that low income communities can be more satisfied with their area than higher income communities (Blom 2009), making the whole community more socially sustainable. In such instances, social sustainability-oriented policies geared towards increasing income in disadvantages communities may be less effective in promoting sustainability than other policies addressing more pressing local social issues. This concern can be also framed within the broader difficulties of developing and applying effective *ex ante* policy assessment frameworks (e.g., identifying sound premises and purposes) as pointed out by Weaver and Jordan (2008).

4 Sustainability Assessment

4.1 Key Features

Sustainability Assessment (SA) is a key element connecting social sustainability research and policy within the context of current calls for evidence-based policies and the appraisal of policies, programmes, plans and projects against sustainability criteria. Broadly speaking, sustainability appraisal is a form of assessment that aims to inform and improve strategic decision making (Sheate et al. 2008). The assessment relies on the application of a variety of methods of enquiry and argument to produce policy-relevant information that is then utilised to evaluate the consequences of human actions against the normative goal of sustainable development (Stagl 2007: 9). Indeed, as Gasparatos et al. (2008) suggest, sustainability assessments ought to:

- Integrate economic, environmental, social and increasingly institutional issues as well as to consider their interdependencies;

- Consider the consequences of present actions well into the future;
- Acknowledge the existence of uncertainties concerning the result of our present actions and act with a precautionary bias;
- Engage the public;
- Include equity considerations (intragenerational and intergenerational).

Sustainability assessment builds on Environmental Impact Assessment (EIA) and Strategic Environmental Assessment (SEA) and despite being a less mature assessment framework than its predecessors, there is general agreement that the assessment is characterised by four main features. These include (i) an emphasis on integration of techniques and themes, (ii) the call for multi-criteria approaches, (iii) the importance of objectives and principles-setting, and (iv) stakeholders' participation in the assessment itself. The in-depth analysis of these aspects is outside the scope of this chapter however, a brief overview of them is provided below in turn.

(i) *Integration of techniques and themes:*

The emphasis in sustainability appraisal is on integration because many approaches to sustainability assessment can be said to be example of 'integrated assessment' (see Weaver and Rotmans 2006, for an extensive analysis of this concept) derived from EIA and SEA, which have been extended to incorporate social and economic considerations as well as environmental ones (Pope et al. 2004; Dalal-Clayton and Sadler 2005). For example, Pope (2007) argues that sustainability assessment can be seen as the 'third generation' of impact assessment processes, following project EIA and the SEA of policies, plans and programmes. From this perspective, EIA-based integrated assessment has been adopted as a sustainability appraisal method by simply replicating the one-dimensional form of assessment in the three-pillar model of sustainable development. This allows for the discrete assessment of the potential environmental, social and economic changes of a proposal and reflects a systemic 'triple bottom line' approach to sustainability (Elkington 1994).

(ii) *Multi-criteria approach*:

There is an increasing call to use a multi-criteria approach in sustainability appraisal in light of the multifaceted nature of the concept that amalgamates social, environmental and economic matters into a new independent entity. For example, in the field of decision making, Multi-Criteria Decision Analysis is an emerging method for sustainability appraisal. It consists of a set of methods using dissimilar criteria, which are combined together by using scores and weightings in order to aid decision making with regards to conflicting evaluations, options and interests. Examples of these methods are Analytic Hierarchy Process, Goal Programming and Novel Approach to Imprecise Assessment and Decision Environments. These appraisal methods acknowledge a pluralist view of society (Glasson et al. 2003) and render the decision-making process more transparent (Stewart 2001). Further, because of the social learning and the reflexive participatory process involved in the assessment, these techniques can help in the evaluation of projects or proposals whose impacts are not well understood and would therefore benefit from a participatory and multi-disciplinary approach (Stagl 2007).

(iii) *Importance of objectives and principles-setting*:

Sustainability appraisal is a form of strategic assessment linked to guiding principles and the achievement of policy objectives. Within this context, Pope et al. (2004) distinguish an objective-led appraisal and a principle-based assessment approach to sustainability. The former is similar in nature to SEA, in which the assessment is carried out to achieve specific policy goals within an explicit framework encompassing environmental, social and economic objectives. The latter is led by objectives derived from broader sustainability principles. In their views, the objective-led appraisal focuses on the appraisal of the 'direction to target', which is usually indicated with '+' '0' or '−' for a positive, neutral and negative move toward the sustainability target. Conversely, the principle-based assessment goes beyond the mere establishment of a 'direction to target' and endeavours to establish the 'distance from target', that is, the extent of progress toward sustainability.

(iv) *Stakeholders' participation in the assessment*:

There has been an increasing call for more participation in the sustainability assessment process (Rotmans et al. 2008) because the latter is often wrongly grounded on the traditional assessor – client relationship (Cavanagh et al. 2007). This form of assessment often fails to understand the varying sensitivity attached to specific issues by a plethora of actors with a stake in the project, process or objective been assessed. Stagl (2007) points out that this traditional technical-rational model of appraisal in which 'objective assessment' by an assessor is assumed to lead automatically to better decisions has proved theoretically, politically and practically inadequate. In his views, the type of assessment can influence its outcome. In other word, the choice of appraisal method and criteria is not a wholly technical question but a 'institutionalising social choice' (Stagl 2007: 3) in which participation is likely to engender a greater sense of ownership of the appraisal process itself (Keough and Blahna 2006).

However, despite the rapid ascent of sustainability assessment techniques in the international arena, the appraisal process has also been subject to criticisms. These include for example,

- Superficiality and lack of quantification (RCEP 2002) of the assessment, which is often due to insufficient provision of benchmarks or the difficulty in establishing how and who should set critical threshold levels for non-environmental variables,

- Stakeholders' involvement is often deemed in practice more consultative rather than participative due to the complexity of the overall assessment process and the availability of resources (Sheate et al. 2008),
- Environmental, economic and social factors are ofter considered separately, with emphasis on balancing the trade-offs between these dimensions rather than exploring the linkages and interdependencies between them (George 2001),
- Lack of consensus concerning the meaning of integrated assessment (Scrase and Sheate 2002) and
- The existence of subjective judgments within the appraisal process concerning integration, win-win solutions and trade-off (Therivel 2004), which make the process not entirely scientific.

These criticisms clearly provide an overview of some of the challenges that will have to be overcome in the field of sustainability assessment, and suggest the main issues that are likely to dominate the sustainability debate in the near future.

4.2 Recent Sustainability Assessment Legislation in the EU

Over the last few decades, sustainability assessment has gained increased recognition in sustainable development policy at the European level, where four main assessment frameworks related to sustainability aspects have been legislated since 1985 (Ruddy and Hilty 2008). These include,

1. Environmental Impact Assessment, which has been typically applied to projects on land use planning at the national level since 1985 through Directives 85/337/EEC and 97/11/EC.
2. Strategic Environmental Assessment came into practice in the mid 1990s as a method to assess the impacts of certain policies, plans and programmes at a higher governance level than land planning. In 2001 the European Council formally adopted the SEA Directive 2001/42/EC that legislates this form of assessment.
3. Sustainability Impact Assessment, introduced by DG trade in 1999 to integrate sustainability into trade policy by informing negotiators of the possible social, environmental and economic consequences of a trade agreement (EC 2005).
4. The EU Impact Assessment System introduced in 2003 by the European Commission to support of the EU's Sustainable Development Strategy and to enhance the quality of the Commission regulatory activity.

If on the one hand, these frameworks demonstrate the variety of assessment techniques legislated at policy level, on the other, they highlight the confusion over the terminology used to measure sustainability and the piecemeal approach that characterises this field. For example, according to the EU terminology, Sustainability Impact Assessment is a process undertaken before and during a trade negotiation in order to identify the economic, social and environmental impacts of a trade agreement (EC 2005). Thus it can be argued that sustainability assessment is currently limited to trade agreements rather than to wider policies, plans and programmes. Furthermore, the methodology developed for the assessment draws upon traditional EIA stages, including Screening – Scoping – Preliminary Assessment – Flanking measures (mitigation and enhancement analysis), but very little is said about the integration criteria and the sustainability principles to be adopted.

To clarify some of the differences and similarities between the main families of assessment techniques, Fig. 2 provides a succinct overview of EIA, SIA, SEA and SA. The diagram offers snapshots of selected definitions, main characteristics and limitations of these forms of assessment. These are meant to summarise rather than replace the very extensive and comprehensive coverage of assessment related issues that can be found in the abundant literature in this field.

4.3 *Conceptual Scope and Practice of Social Sustainability Assessment*

From a social sustainability perspective, there is paucity of specific sustainability assessment methodologies as such. The assessment is often conducted through

Increasing integration, strategicness and comprehensiveness of themes and methods

	Since 1960s EIA	1970s SIA	1990s SEA	2000s SA
Selected definitions and objectives	A public process by which the likely effects of a project on the environment are identified, assessed and then taken into account by the consenting authority in the decision-making process	A systematic, iterative, ex-ante form of assessment that seeks help individuals, groups, organizations and communities understand possible social and cultural, or economic impacts of change, or better still impacts of proposed change	A form of environmental assessment intended to identify and assess the likely significant effects of a plan, programme or a policy on the environment, the result of which are then taken into account in the decision-making process	A form of strategic assessment that integrates environmental, social and economic parameters and relies on the application of a variety of methods of enquiry and argument to produce policy-relevant information in order to evaluate human actions against the normative goals of sustainable development
Main Features	• Focus on environmental dimension of sustainable development, though it may include separate social considerations • Physical/Quantitative approach to the measurement of selected variables • Selection of objective but contextual targets and thresholds • Limited to project level	• Focus on social dimension • Speculative in nature, does not provide precise, accurate and repeatable results • The selection of targets and thresholds relies on system values and political objectives rather than scientific criteria • Primary, secondary, cumulative and 'dead-weight' effects are difficult to calculate and measure	• operates at a strategic level • stresses process rather than detailed technical analysis • foundations in EIA but by nature more open-ended, consultative and iterative than EIA • No need for sophisticated and expensive data gathering and modelling capacity • inter-institutional cooperation and public participation key determinants of success	• Integration of sustainable development dimensions • relies upon principles and objectives rather than targets and thresholds • acknowledge the existence of uncertainties concerning the result of our present actions and act with a precautionary bias • engage the public • include equity considerations (intra generational and intergenerational).
Examples of main limitations	• Ignores politics and models of decision making • Too narrow focus on bio-physical environment	• Quality and availability of data at the local level • 'Social engineering' risk	• Environmental effects hard to predict at strategic level • Achieving integration	•Quantification issues •Trade-offs, aggregation and weights difficulties

EIA = Environmental Impact Assessment; SIA = Social Impact Assessment; SEA: Strategic Environmental Assessment; SA = Sustainability Assessment

Fig. 2 Overview of main methods to assess sustainable development and its dimensions (*Source*: Various. See also Colantonio (2008))

social impact assessment (SIA), which is extended to include other sustainability pillars. For example Hacking and Guthrie (2008) maintain that the extended coverage of sustainability appraisal is being accommodated by 'stretching' EIA or SEA and broadening the definition of 'environment' and hence the thematic coverage of theme-specific assessment such as SIA. However, they question the real level of integration of these techniques because in their views SIA may be undertaken on its own, as a component of EIA, in parallel with EIA, or as part of an 'integrated' S&EIA. It is also worth pointing out that these diverse impact assessment techniques were not designed for sustainability appraisal *per se*. As a result, their semantic or substantive integration may not be able to capture, address and suggest solutions for a diverse set of issues that affect stakeholders with different values and span over different spatial and temporal scales (Gasparatos et al. 2007).

Within this context, in a recent study of 20 Environmental Statements (ESs) concerning randomly selected urban regeneration projects implemented in the UK between 1998 and 2007, Glasson and Wood (2008) point out that SIA is covered in 80% of the cases, often in a separate chapter. According to their analysis, the scope of SIA content has widened from the 1990s experience to cover population profile and occupational groups; economic and business context; learning and employment; general well being, health, crime and deprivation; community facilities and services; recreation and public open space; and social inclusion and community integration. Further, they argue that there is increasing evidence of best practices in project-SIA after 2004, partly because of the publication of the Planning and Compulsory Purchase Act (United Kingdom Government 2004) and the Sustainability Appraisal of Regional Spatial Strategies and Local Development Document (ODPM 2005).

However, they also note that there is limited evidence of a sustainability approach that set the SIA and ESs within a wider sustainability context. This is for example because (i) only 50% of ESs contain methodological information that goes beyond a bland descriptive review of population and employment baseline (ii) there is insufficient analysis of the links between socio-economic components (e.g., between demographic profile and jobs created), (iii) quantification is limited and mainly focused on demographics, employment, services and facilities provision, and (iv) the assessment methods showed limited community engagement and reduced involvement of a wide range of stakeholders.

Lastly, at a more conceptual level it can also be argued that another fundamental problem for the deployment of SIA within a sustainability perspective concerns the target and threshold-setting exercise inherent to the impact assessment itself, which presents problems when applied to social settings. Indeed, the bad experience of the 1960s makes social scientists hesitant to formulate normative targets and thresholds, and there can be little doubt that social engineering policies of the 1960s have been criticised for promoting ill-conceived social formulations (Omann and Spangenberg 2002 In addition, social objectives against which to assess social sustainability need to be contextualised within different development models and system values. These range from neoliberalism policies to the European social security model and to more eclectic approaches to development adopted by transitional economies and continuing socialist countries.

4.4 Social Sustainability Metrics

Historically, long lists of indicators were established to describe the complexity of sustainable development, with special focus on its environmental dimension. A recent study by Therivel (2004) showed that two thirds of sustainability indicators addressed environmental concerns. More recently, these rather technical lists have been enlarged to include social indicators. Long lists have also been simplified and reduced to sets of core indicators (Hens and De Wit De 2003), which are 'bundled' into sustainability themes, objectives and guiding principles. These elements are interlinked together and constitute the backbone of most sustainable development policies.

In terms of social sustainability metrics, previous work from Colantonio (2007) argued that

- The evolution of indicators shows how older indexes prioritise the basic needs component whilst indicators developed more recently seem to emphasise the importance of governance, representation and other institutional factors (see Colantonio 2007 for a review of this evolution).
- In older indexes the elements taken into account were technically weighted together with other dimensions of sustainable development in an attempt to deliver an integrated approach to sustainability. However, in later sustainability indicators the final decision about trade-offs is de facto left to 'sound judgement', as well as leadership and communication skills (Egan 2004).
- The 'community' and the 'local level' have re-emerged as main spatial and operational space for the pursuit of sustainability.
- There has been a shift from purely statistics-based indicators toward hybrid sets of indicators that mix quantitative data and qualitative information.

Broadly speaking, the review of recent developments and trends in social sustainability assessment and measurement also suggests a broad distinction between 'traditional social indicators' and 'social sustainability indicators', which is summarised in Table 4. According to this categorisation, it can be argued that traditional social indicators are used for the analysis of discrete issues accessible to specific methodologies related to individual themes that are linked to targets rather than objectives. They are also often selected by panels of experts in national and regional statistical offices. They focus on targets or outcomes and provide a static analysis of national and regional social phenomena.

By contrast, social sustainability indicators are concerned with the integration of multidimensional and intergenerational issues inherent to the notion of sustainability. Their selection is informed by sustainability principles and objectives, which stem

Table 4 Characteristics of traditional social indicators and social sustainability indicators

Traditional social indicators	[Emerging] social sustainability indicators
Static	Intergenerational and incorporating uncertainty
Predominantly quantitative	Hybrid
Product	Process
Descriptive	Strategic
Mono-dimensional	Multi-dimensional
Target oriented	Principles and objective driven
Top down selection	Deliberative and reiterative selection

from a deliberative and reiterative participation process involving a wide array of stakeholders and local agents. Moreover, sustainability indicators are *process indicators* in the sense that they analyse the processes through which sustainability principles and objectives are defined, themes agreed and solutions implemented. They allow the monitoring of the actual implementation of a project or a phenomenon and assess the progress towards specific objectives in a more interactive way than traditional social indicators.

To briefly clarify and exemplify these differences we can look, for example, at how poverty would be 'measured' from a 'traditional perspective' as opposed to a 'social sustainability perspective'. The traditional approach to measuring poverty involves establishing an income threshold and calculating how many individuals, families or households fall below it (Townsend and Kennedy 2004). Poverty is measured in a discrete way and linked for instance to a poverty reduction target. By contrast, from a sustainability perspective, poverty would be measured together with its main manifestations – e.g., ill-health, inadequate housing, limited access to basic services etc. – in a multi-dimensional index that integrates the processes and factors conducive of poverty. These include for example marginalisation, inability to access to education etc.

From an operational perspective, however, the aggregation of singles indexes and dimensions presents several difficulties. For example, current integrative frameworks still do not allow a meaningful aggregation of diverse metrics. Keirstead (2007), for instance, comments that it is not clear how data of fuel poverty and quality of life can be combined into a single social sustainability metric. Even if data can be normalised and weighted, it proves difficult to aggregate social, environmental, economic and institutional metrics into a composite index that can be compared at both spatial and temporal levels.

The development and integration process of indicators is hindered further by the shift in the social sustainability discourse from the in-depth analysis of hard themes towards the inclusion of soft themes, as reviewed earlier. As a result, new sustainability indicators are increasingly focused on measuring these emerging themes rather than improving the measurement of more traditional concepts such as equity and fairness. For example, if on the one hand, a growing number of variables and factors are being proposed to deconstruct and measure happiness and well being of individuals and communities worldwide (Veenhoven and Hagerty 2006), on the other, the main approach to equity still relies on the analysis of income and relative prosperity, as shown for example by recommendations contained in the UK Green Book (Treasury 2005), a recent guideline document for the appraisal of governmental policies, plans and projects.

Recent sets of sustainable development indicators also illustrate the tendency of favouring the investigation of softer themes at the expenses of sophisticating the measurement of more established social sustainability pillars. For instance the latest set of sustainable development indicators released by the UK government in 2007 (ONS and DEFRA 2007) contains a Sustainable Communities and a Fairer World cluster of indicators, addressing social sustainability concerns. This cluster suggests several indicators to assess different aspects of sustainable communities, including well-being, life satisfaction etc. However, it does not recommend any index to deal with the interlinked subjects of social justice, equity, fairness, and cohesion (ONS and DEFRA 2007: 96). Similarly, a recent study commissioned by the EU Parliament (EP 2007) to look at the implementation of the Sustainable Communities approach in the EU concluded that fairness cannot be adequately measured through existing indicators and further work is needed in this area.

5 An Example of Emerging Social Sustainability Indicators: The Sociale Index in Rotterdam

Despite the theoretical and practical hurdles to monitor the social evolution of places, this section illustrates the practical example of the 'Sociale Index' (see Fig. 3), which is a composite index launched in 2008 by Rotterdam municipal authorities to monitor the transformation of the social qualities of the city's neighbourhoods.

The index collects and aggregates data concerning four main dimensions of Rotterdam's areas and their residents, including (i) ***personal abilities*** (language skills, health, income, education), (ii) ***living environment*** (level of discrimination, housing, public facilities, safety, etc.), (iii) ***participation*** (going to work/school, social contact, social and cultural activities, etc.), and (iv) ***'bonding'*** (mobility, 'feeling connected', etc.) (Leidelmeijer et al. 2007). The index produces a score between 0 and 10, which has four main purposes (Koppelaar 2009; Leidelmeijer et al. 2007; Rotterdam 2008), including:

- Measuring the social qualities of a place at a given time;
- Showing and comparing the differences between 64 of the 80 districts of Rotterdam;
- Providing a baseline for the assessment of policies;
- Analysing the strengths and weaknesses of each neighbourhood in terms of the dimensions included in the index.

The index, which is calculated yearly, comprises statistical (30%) and survey (70%) data.

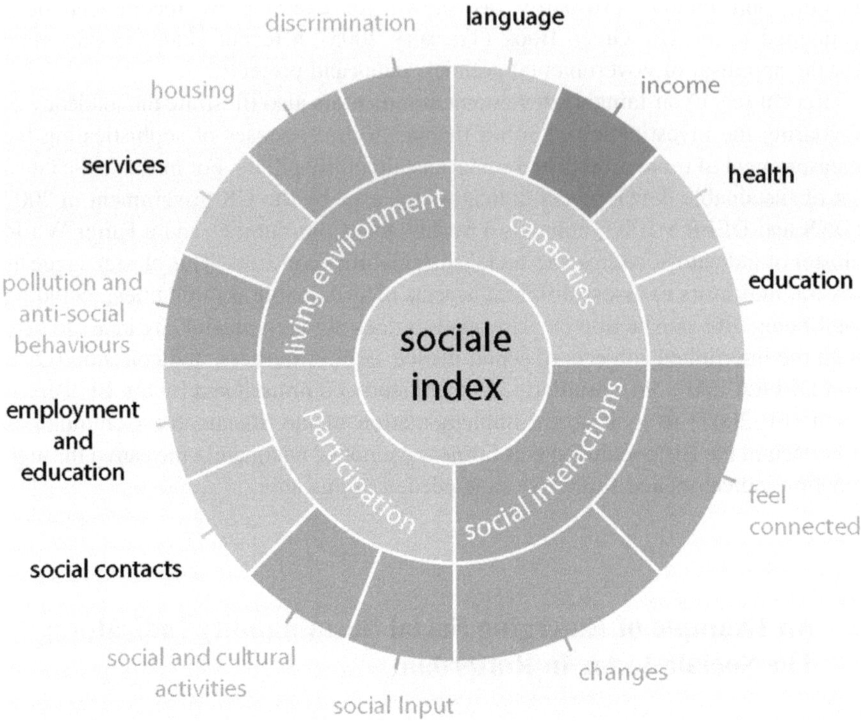

Fig. 3 The sociale index (*Source*: Translated and re-drawn by Colantonio from Rotterdam (2008))

The Sociale Index is relatively new, and no comprehensive evaluation of the pros and cons of this index have been carried out thus far. There can be little doubt, however, that this index embodies some of the main characteristics of emerging sustainability indicators reviewed in Table 4. Indeed, the Sociale Index is a multi-dimensional and hybrid indicator in the sense that it endeavours to aggregate different social sustainability themes together through a mix of qualitative (survey to measure participation and bonding) and quantitative data (official statistics for living environment and personal abilities) analysis. In addition, data for the calculation of the sociale index is gathered at neighbourhood level in order to provide an overview of how people live together, participate in local community activities and feel connected with each other (RIGO 2007).

For the purpose of this chapter this innovative indicator will not be examined further from an empirical point of view, that is, through the investigation of its operational and practical implications. However, it can be argued that, if read in conjunction with previous sections of this chapter, the brief analysis of the Sociale Index highlights the main methodological and theoretical issues involved in the measurement of social sustainability (at city level). Further it provides an opportunity for policy-makers and practitioners alike to reflect on the key rationale and methods that should be adopted to conceptualise the evaluation of local, national and international social policies.

6 Conclusions

This chapter has provided a concise overview of the social dimension of sustainable development and suggested a taxonomical division between traditional and emergent social sustainability themes and indicators. This has been instrumental to suggest that the shift toward the analysis of more elusive concepts in the social sustainability debate may continue for the foreseeable future as larger sectors of communities and societies become more affluent and less worried about the satisfaction of basic needs. It is important however that this new focus on emerging themes is not pursed at the expense of more in-depth analysis of traditional pillars of social sustainability, such as equity and poverty, which have received less attention in recent social sustainability works.

The chapter has also illustrated how the progress toward sustainability is increasingly being appraised by extending and integrating ‘Impact Assessment’ and ‘Strategic Impact Assessment’ methods into ‘Sustainability Assessment’. Techniques such as Environmental Impact Assessment, Strategic Environmental Assessment, Social Impact Assessment, Health Impact Assessment etc. are being amalgamated into a new independent form of assessment rooted in the philosophical and methodological framework provided by sustainability. However, these early forms of impact assessment were not designed to address the complexity inherent to the measurement of sustainability. As a result, there is widespread uncertainty concerning for example how different typologies of impact and assessment techniques should be integrated together.

Future research will also have to focus on unravelling the underlying inter- and intra-linkages between social sustainability assessment methods, policies, indicators and themes (for example equity and happiness or well-being and identity etc.). Further, it will have to investigate how the latter can be ‘quantified’ using simple and user friendly methods capable of deconstructing and monitoring these elements without losing the richness of information that is embedded within them. Lastly, another major challenge for future social sustainability research in the EU is to carry out systematic studies to understand the mechanisms, channels, processes, actors and timeline through which innovative ideas and theories are integrated in EU sustainable development policies and legislation.

Acknowledgments I would like to thank Prof. Tim Dixon for his numerous helpful editorial comments, which have significantly improved the structure and the content of this chapter through various drafts. I also wish to thank Prof. John Glasson for his valuable suggestions especially concerning the sections examining sustainability assessment and indicators.

References

Ancell, S., & Thompson-Fawcett, M. (2008). The social sustainability of medium density housing: A conceptual model and Christchurch case study. *Housing Studies, 23*(3), 423–442.
Assefa, G., & Frostell, B. (2007). Social sustainability and social acceptance in technology assessment: A case study of energy technologies. *Technologies in Society, 29*, 63–78.
Baines, J., & Morgan, B. (2004). Sustainability appraisal: A social perspective. In B. Dalal-Clayton & B. Sadler (Eds.), *Sustainability appraisal. A review of international experience and practice.* London: First Draft of Work in Progress, International Institute for Environment and Development.
Biart, M. (2002) Social sustainability as part of the social agenda of the European community. In: *Soziale Nachhaltigkeit: Von der Umweltpolitik zur Nachhaltigkeit?* (ed. T. Ritt), Arbeiterkammer Wien, Informationen zur Umweltpolitik 149, Wien, pp. 5–10. (Retrieved June 2008 from http://wien.arbeiterkammer.at/pictures/importiert/Tagungsband_149.pdf)
Blom, D. (2009, February 19–20). *Pact op Zuid – Pact of South.* Presentation at workshop on social sustainability and urban regeneration, Oxford Brookes University, Oxford.
Bramley, G., Dempsey, N., Power, S., & Brown, C. (2006). *What is 'Social Sustainability' and how do our existing urban forms perform in nurturing it?* Paper presented at the 'Sustainable Communities and Green Futures' conference, Bartlett School of Planning, University College London, London.
Brundtland Commission (1987) Our Common Future. World Commission on Environment and Development, New York
Cavanagh, J. A., Frame, B. R., Fraser, M., & Gabe, G. (2007, June 27–29). *Experiences of applying a sustainability assessment model.* International conference on whole life urban sustainability and its assessment, SUE-MoT conference proceedings, Glasgow.
Chambers, R. (1992). *Rural appraisal: Rapid, relaxed and participatory* (IDS discussion paper 311, pp. 69–84). Brighton: IDS.
Chambers, R., & Conway, G. (1992) *Sustainable Rural Livelihoods: Practical Concepts for the 21st Century*, IDS Discussion Paper 296. IDS, Brighton.
Chiu, R.L.H. (2003) Social sustainability, sustainable development and housing development: The experience of Hong Kong. In: *Housing and Social Change: East–West perspectives* (eds, R. Forrest & J. Lee), pp. 221–239. Routledge, London.
Colantonio, A. (2007). *Social sustainability: An exploratory analysis of its definition, assessment methods* (Metrics and Tools, OISD (EIB) Working Paper 2007/01). Oxford: Oxford Institute for Sustainable Development.
Colantonio, A. (2008). *Traditional and emerging prospects in social sustainability* (OISD (EIB) Working Paper 2008/03). Oxford: Oxford Institute for Sustainable Development.
Coleman, J. S. (1988). Social capital and the creation of human capital. *The American Journal of Sociology, 94*(Supplement), S95–S120.
Dalal-Clayton, B., & Sadler, B. (2005). *Sustainability appraisal: A review of international experience and practice*. London: Earthscan Publications.
DFID (1999) Sustainable Livelihoods Guidance Sheets. Department for International Development, London.
Drakakis-Smith, D. (1995) Third world cities: Sustainable urban development, 1. Urban Studies, 32(4–5), 659–677.
Egan, J. (2004) The Egan Review: Skills for Sustainable Communities. ODPM, London.
Elkington, J. (1994). Towards the sustainable corporation: Win-win-win business strategies for sustainable development. *California Management Review, 36*(2), 90–100.
European Commission (EC). (2005). Sustainability impact assessment. Available at: http://ec.europa.eu/trade/issues/global/sia/faqs.htm. Accssed in May 2008.
European Parliament (EP). (2007). *The possibilities for success of the sustainable communities approach and its implementation*, European Parliament Study Directorate-General for Internal Policies of the Union Structural and Cohesion Policies Policy Department.

Gasparatos, A., El-Haram, M., & Horner, M. (2007). *The argument against a reductionist approach for assessing sustainability*, paper presented at SUE-MOT International Conference on Whole Life Urban Sustainability and its Assessment, Glasgow Caledonian University, Glasgow, 27th–29th June 2007

Gasparatos, A., El-Haram, M., & Horner, M. (2008) A critical review of reductionist approaches for assessing the progress towards sustainability. Environmental Impact Assessment Review 28(4–5): 286–311

George, C. (2001). Sustainability appraisal for sustainable development: Integrating everything from jobs to climate change. *Impact Assessment and Project Appraisal, 19*(1), 95–106.

Glasson, J., & Wood, G. (2008). *Urban regeneration and impact assessment for social sustainability*. Paper presented at the IAIA08 conference, Perth.

Glasson, J., Therivel, R., & Chadwick, A. (2003). *Introduction to environmental impact assessment: Principles and procedures, process, practice and prospects*. London/Philadelphia: UCL Press.

Great Vancouver Regional District (GVRD). (2004). *The social components of community sustainability: A framework*. Vancouver: TAC Social Issues Subcommittee.

Hacking, T., & Guthrie, P. (2008). A framework for clarifying the meaning of triple bottom-line, integrated, and sustainability assessment. *Environmental Impact Assessment Review, 28*(2–3), 73–89.

Hans-Böckler-Foundation (ed.) (2001) *Pathways Towards a Sustainable Future*. Setzkasten, Düsseldorf.

Hardoy, J., Mitlin, D., & Satthertwaite, D. (1992). *Environmental problems in third world cities*. London: Earthscan Publications.

Healey, P. (1992). Planning through debate: The communicative turn in planning theory. *Town Planning Review, 63*(2), 143–162.

Hens, L., & De Wit De, J. (2003). The development of indicators for sustainable development: A state of the art review. *International Journal of Sustainable Development, 6*, 436–459.

Keirstead, J. (2007) *Selecting Sustainability Indicators for Urban Energy Systems*. International Conference on Whole Life Urban Sustainability and its Assessment, SUEMoT Conference Proceedings Glasgow, 27–29 June.

Keough, H. L., & Blahna, D. J. (2006). Achieving integrative, collaborative ecosystem management. *Conservation Biology, 20*(5), 1373–1382.

Koppelaar, P. (2009, February 19–20). *'Sociale index: A social monitor for the municipality of Rotterdam'*. Presentation at the urban regeneration and social sustainability workshop, Oxford Institute for Sustainable Development, Oxford Brookes University.

Layard, R. (2005). *Happiness: Lessons from a new science*. New York: Penguin.

Layard, R. (2007). Against unhappiness. *Prospect* (on line-version), 137.

Leidelmeijer K., van Iersel J., & den Herder m.m.v. N. (2007). *Sociale index Rotterdam Bijlagenrapport*. Unpublished, RIGO Research en Advies BV, Amsterdam.

Littig, B., & Grießler, E. (2005) Social sustainability: A catchword between political pragmatism and social theory. International Journal of Sustainable Development, 8(1–2), 65–79.

Loorbach, D., & Rotmans, J. (2006). Managing transitions for sustainable development. In X. Olshoorn & A. J. Wieczorek (Eds.), *Understanding industrial transformation: Views from different disciplines*. Dordrecht: Springer.

Metzner, A. (2000). Caring Capacity and Carrying Capacity – A Social Science Perspective, Paper presented at the INES 2000 Conference: Challenges for Science and Engineering in the 21st Century, Stockholm.

Nussbaum, M., & Glover, J. (Eds.). (1995). *Women, culture, and development: A study of human capabilities* (pp. 360–395). New York: Oxford University Press.

OECD. (2001). *Analytic report on sustainable development SG/SD(2001)1-14*, Paris: OECD.

Office of National Statistics (ONS) and Department for Environment Food and Rural Affairs, (DEFRA). (2007). *Sustainable development indicators in your pocket*. London: Defra Publications.

Office of the Deputy Prime Minister, ODPM. (2005). *Sustainability appraisal of regional spatial strategies and local development documents*. London: ODPM.

Omann, I., & Spangenberg, J.H. (2002, March 6–9). *Assessing social sustainability. The social dimension of sustainability in a socio-economic scenario*. Paper presented at the 7th biennial conference of the international society for ecological economics in Sousse, Tunisia.

Ormerod, P., & Johns, H. (2007). Against happiness. *Prospect* (online version), 133.

Polese, M. & Stren, R. (eds) (2000) *The Social Sustainability of Cities: Diversity and the Management of Change*. University of Toronto Press, Toronto, Canada.

Pope, J. (2007, January 4–7). *Sustainability assessment as a deliberative learning process*. Presentation at sustainability conference, University of Madras, Chennai.

Pope, J., Annandale, D., & Morrison-Saunders, A. (2004). Conceptualising sustainability assessment. *Environmental Impact Assessment Review, 24*, 595–616.

Putnam, R. D. (1993). *Making democracy work: Civic tradition in modern Italy*. Princeton: Princeton University Press.

RIGO. (2007), *Sociale index Rotterdam Bijlagenrapport*. Amsterdam: RIGO Research en Advies BV.

Rotmans, J., Kemp, R., & van Asselt, M. (2001). More evolution than revolution: Transition management in public policy. *Foresight, 3*(1), 1–17.

Rotmans, J., Jäger, J., & Weaver, P. M. (2008). Editorial. *International Journal of Innovation and Sustainable Development, 3*(1/2), 1–8.

Rotterdam, G. (2008). *Rotterdam sociaal gemeten.1e meting door de Sociale Index*. Rotterdam: City of Rotterdam Publishing.

Royal Commission on Environmental Pollution (RCEP). (2002). *23rd Report on environmental planning, Cm 5459*. London: The Stationery Office.

Ruddy, T. F., & Hilty, M. L. (2008). Impact assessment and policy learning in the European Commission. *Environmental Impact Assessment Review, 28*(2–3), 90–105.

Sachs, I. (1999). Social sustainability and whole development: Exploring the dimensions of sustainable development. In B. Egon & J. Thomas (Eds.), *Sustainability and the social sciences: A cross-disciplinary approach to integrating environmental considerations into theoretical reorientation*. London: Zed Books.

Sachs, J. (2001). The strategic significance of global inequality. *The Washington Quarterly, 24*(3), 187–198.

Scrase, J. I., & Sheate, W. R. (2002). Integration and integrated approaches to assessment: What do they mean for the environment? *Journal of Environmental Policy & Planning, 4*, 275–294.

Sen, A. K. (1985). *Commodities and capabilities*. Oxford: Oxford University Press.

Sen, A. K. (1992). *Inequality re-examined*. Oxford: Clarendon.

Sen, A. (1993). Capability and well-being. In M. Nussbaum & A. Sen (Eds.), *The quality of life* (pp. 30–53). Oxford: Oxford University Press.

Sheate, W. R., Rosario do Partidario, M., Byron, H., Bina, O., & Dagg, S. (2008). Sustainability assessment of future scenarios: Methodology and application to mountain areas of Europe. *Environmental Management, 41*, 282–299.

Sinner, J., Baines, J., Crengle, H., Salmon, G., Fenemor, A., & Tipa, G. (2004). *Sustainable Development: A summary of key concepts*. Ecologic Research Report No. 2, available at www.ecologic.org.nz, accessed in May 2009

Stagl, S. (2007). *Emerging methods for sustainability valuation and appraisal – SDRN rapid research and evidence review* (p. 66). London: Sustainable Development Research Network.

Stewart, M. (2001). *MMSD life cycle assessment workshop: the application of life cycle assessment to mining, minerals and metals*. London: Centre for Risk, Environment and Systems Technology and Analysis (CRESTA) and Department of Chemical Engineering. University of Sydney for the International Institute for Environment and Development (IIED).

Therivel, R. (2004). *Sustainable urban environment-metrics models and toolkits-analysis of sustainability/social tools*. Oxford: Levett-Therivel.

Thin, N., Lockhart, C., & Yaron, G. (2002). *Conceptualising socially sustainable development*. A paper prepared for DFID and the World Bank, DFID, Mimeo, Washington, DC.

Townsend, I., & Kennedy, S. (2004). *Poverty: Measures and targets* (Research Paper 04/23). London: Economic Policy and Statistics Section, House of Commons Library.
Treasury, H. M. (2005). *The green book, treasury guidance*. London: TSO.
United Kingdom Government. (2004). Planning and Compulsory Purchase Act, London.
Veenhoven, R., & Hagerty, M. (2006). Rising happiness in nations 1946–2004. *Social Indicators Research, 79*, 421–436.
Wackernagel, W., & Rees, M. (1996). *Our ecological footprint: Reducing human impact on the earth*. Gabriola Island, BC and New Haven: New Society Publisher.
Weaver, P. M., & Jordan, A. (2008). What roles are there for sustainability assessment in the policy process? *International Journal of Innovation and Sustainable Development, 3*(1/2), 9–32.
Weaver, P. M., & Rotmans, J. (2006). Integrated sustainability assessment: What is it, why do it and how? *International Journal of Innovation and Sustainable Development, 1*(4), 284–303.

Dealing with Doom: Tackling the Triple Challenge of Energy Scarcity, Climate Change and Global Inequity

Jos Bruggink

Preamble. On December 1, 2008, flood waters inundated Venice once more, reaching a 20-year high of 156 cm above normal. The event scared tourists a lot more than Venetians. They have learned to adapt. Of course, Venetian authorities have seen it coming for 20 years. A massive flood barrier has been planned for decades. It was due to become operational in 2011, but financial problems once again have delayed installation by a perennial couple of years. The flood barrier is known by the apt biblical acronym MOSES. Venice is full of biblical reminders and works of art referring to doom and salvation abound. In fact, Venice has been a city of doom and salvation long before the floods started to become serious. It lost its position as a global power two centuries ago. It lost two thirds of its population in the past century and now it seems to be sinking for good. No wonder that many books have been written on the death of Venice. Yet, the city survives. Visitors keep coming in droves and Venice seems to be a place of merry wealth rather than sad poverty. The city has been courting doom for ages, yet it seems splendidly sustainable in the sense of being able to adapt resiliently. They turned doom into their own brand of sustainability. The fate of Venice invites reflection on the essence of doom and thereby the essence of sustainability. It provides an admittedly crude, yet appropriately warning, symbol for Europe's future. The Venetian story is a harsh reminder, that Europe must develop its own brand of sustainability. The Venetian story also tells that sustainability may have more to do with surviving doom smartly than with attaining salvation purposefully. How doom and sustainability are intimately connected, and what this implies for a European brand of sustainability, are in fact the major themes of this analysis of the energy related faces of doom.

J. Bruggink (✉)
Unit Policy Studies, Energy research Centre of the Netherlands (ECN), Westerduinweg 3, 1755 LE Petten and Institute for Environmental Studies, Vrije Universiteit (VU), Amsterdam, The Netherlands
e-mail: bruggink@ecn.nl

C.C. Jaeger et al. (eds.), *European Research on Sustainable Development*,
DOI 10.1007/978-3-642-19202-9_6,
Published by Springer-Verlag Berlin Heidelberg 2011

1 Sustainability Perspectives, Policy Strategies and Issue Linkages

1.1 Scientific Perspectives on Global Sustainability

The literature is full of interpretations of the concept of sustainability. Three different scientific perspectives can be distinguished: a normative-prescriptive perspective, an analytical-empirical perspective and a strategic action-oriented perspective. The first, normative-prescriptive perspective basically addresses the question of what a sustainable world should look like. It does not answer the question of how to reach such a world effectively. Normative-prescriptive perspectives concern the delineation of appropriate ambitions. The second, analytical-empirical perspective addresses the question of how to develop concrete indicators for sustainability and how to measure those empirically. It does not answer the question of how to set ambitions for those indicators. It primarily addresses the question of what can be monitored scientifically about sustainability however defined. Finally, the strategic action-oriented approach is focused on a policy perspective and answers the pragmatic question of what can be done effectively now to improve the prospects for sustainability later even if we are unsure of, do not agree on or are unable to measure its desirable features precisely. Strategic action-oriented approaches have to do with the design and implementation of policy measures. In this analysis sustainability is addressed exclusively from the third, strategic action-oriented perspective.

1.2 First Policy Principle: Balance Between Shaping and Hedging Strategies

Policy professionals in the field of energy and sustainability are generally an optimistic lot. They tend to focus on the happy and vague contours of a sustainable future rather than the unfortunate and clear progress the world is making towards an unsustainable future now. Many discussions on sustainability focus on the urgency of future ambitions rather than the predicaments of the recent past. This focus on ambitions for sustainability presupposes a manageable world, where Europe is actually able to promote sustainability effectively at the global level. Policy strategies presupposing a manageable world that can be shaped by public action, I will call shaping strategies. A shaping strategy is driven by a "nothing ventured, nothing gained" policy attitude. In a shaping strategy, R&D policies rather than implementation policies form a key element of strategy, because in an evolutionary sense there is need to create diversity for technological innovation rather than to choose a winning technology prematurely. A shaping strategic approach can be contrasted with a hedging strategic approach. Such an approach presupposes that we are living in a dangerous and uncertain world, where we have few opportunities to make a lasting impact. Hedging strategies address an unmanageable world where it would be wise to guard against the unfortunate consequences of unsustainable ambitions rather than to persist in the illusion of shaping sustainable futures.[1] A hedging strategy is driven by a "better be safe than sorry" policy attitude. In a hedging strategy, implementation policies rather than R&D policies form a key element of strategy, because in an evolutionary sense there is a need to create a stable selection environment and choose winning technologies at an early stage instead of keeping options open permanently. Hedging strategies primarily aim to avoid doom, while shaping strategies primarily aim to attain salvation.[2] My first major proposition for effective design of policies in Europe is that a balanced and interconnected combination of shaping strategies and hedging strategies will make policies for sustainability much more effective.

1.3 Second Policy Principle: Hierarchical Ordering of Three Faces of Energy Doom

Doom in terms of energy-related sustainability problems has three major faces: impending energy scarcity, creeping climate disasters and persistent energy poverty. Scientists in the sustainability arena tend to frame these issues in a world view that can be pictured as a triangle of the three p's: profit (economic objectives), planet (environmental objectives) and people (social objectives) with each corner representing a separate, but equally important aspect of sustainability. The three faces of energy doom correspond to these three corners of the sustainability triangle (see the left part of Fig. 1). The depletion of hydrocarbon resources threatens economic objectives, the pace of climate change threatens environmental objectives and the lack of access to energy for the poor threatens social objectives. This type of triangular framing of global sustainability problems has a major limitation: it tends to neglect the hierarchical priorities attached to sustainability objectives in a world characterized by increasing geopolitical complexity and a bewildering multiplication of global actors. In this multipolar world of increasing strategic uncertainties and risks, the sustainability triangle should be framed as a hierarchy of objectives, more or less in analogy with the Maslov pyramid of

[1]Not unsurprisingly, the military intelligence community is keen on considering the consequences of living in a highly dangerous and uncertain world. Their scenarios are a welcome reminder of the geopolitical realities in which a transition towards global sustainability must be made. See for instance the report of the US National Intelligence Council (NIC 2008).

[2]The terms shaping and hedging strategies in the context of long-term energy futures were first used in a Rand corporation study on US energy scenarios. See (Silberglitt et al. 2003).

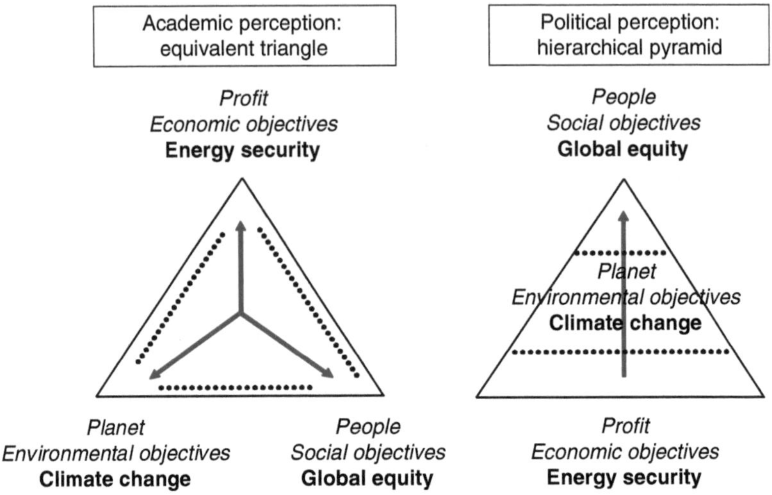

Fig. 1 Perceptions of the concept of sustainability

individual needs (see the right part of Fig. 1). I emphasize that this hierarchical ordering is based on adopting a strategic-action oriented perspective, not on an inherently accepted ethical position.[3] Put more harshly, we simply live in a world where global governance structures are somewhat able to deal with economic objectives, weakly able to deal with environmental objectives and not at all able to deal with social objectives. This pragmatic observation has a lot to do with the present global governance structure that, apart from military might, has so far been characterized by almost total reliance on the market. In consequence, my second major principle for effective design of policies in Europe is that approaching the three faces of energy doom sequentially and in hierarchical order will make policies for sustainability much more effective.

1.4 Third Policy Principle: Linking Sustainability Issues Dynamically

In practice, the professional and policy communities active in each corner of the sustainability triangle tend to operate in relative isolation, both institutionally and intellectually. Moreover, it would be fair to state that Europe has been most

[3]The analogy between a hierarchy in energy policy objectives and a hierarchy in individual personal motivations (the Maslow pyramid) was first used by Chistoph Frei in 2004. See (Frei 2004).

effective in developing common and cohesive policies in the area of climate change mitigation while progression in the areas of energy security and global poverty is comparatively slow. Climate change problems occupy the central stage in the European energy policy arena, although it should be emphasized that this observation concerns the mitigation part of climate change problems, not the adaptation part. Problems of resource depletion and persistent poverty are generally viewed as less tractable, both analytically and politically. This climate change oriented focus in European energy policy tends to neglect the dynamic nature and changing importance of issue linkages between the three faces of doom. Moreover, there is a tendency to aim for speeding up implementation of climate change policies prematurely rather than given due emphasis to R&D first. In contrast to present perceptions of policy priorities, I will argue that climate change mitigation policies are likely to become effective if and only if energy scarcity problems become more urgent and visible, and that persistent global energy poverty can be addressed more effectively if and only if climate change problems become more urgent and visible. Therefore, my third principle for effective energy policies in Europe is that policy makers have to link the issues of climate change with the issues of energy scarcity and global inequity in a much more subtle and phased way than is often presently the case.

2 Vision on European Energy Policy Strategies

2.1 Storyline for a European Brand of Energy Sustainability

The strategic policy principles outlined above can be combined and used to sketch a storyline for an effective European energy policy in the coming decades. This storyline unfolds in three stages, each representing a different ordering of the three priorities for energy strategy. The first phase in the storyline involves a strong emphasis on hedging policies to address energy security coupled with a strong emphasis on shaping policies to address climate change. Issues of global equity are only marginally addressed at this stage. The second phase in the storyline involves a strong emphasis on hedging policies to address climate change coupled with a strong emphasis on shaping policies to address global inequity. Issues of energy security are less prominent at this stage. Finally, the last phase in the storyline involves a balanced package of policies to address the three energy-related sustainability problems mentioned simultaneously. In this final phase consolidation of the new sustainable energy regime takes place, while the first two stages are typical transitional phases. The timeline for these three phases is not fixed, but depends on global developments outside the scope of European policy influence, so the three phases may turn out to be of different length.

2.2 *Rationale for Three Phases in European Energy Policy Strategy*

The storyline for a European brand of energy sustainability assumes that the immediate threat for Europe in the coming decade is the scarcity of hydrocarbon resources and that the arrival of recurrent supply shortages and dramatic price increases is likely to dominate political priorities in the near future.[4] The second phase in this storyline kicks in when the immediate threat shifts from hydrocarbon security concerns to climate change security concerns and the necessity and costs of adaptation become much more clear. This threat will become apparent by world-wide ecological crises in the medium term that will affect global economic performance and will move climate change policies high on the global political agenda. By that time fossil fuel prices will have increased substantially. This will level the economic battleground between energy technologies sufficiently to give climate-friendly alternatives a chance to compete effectively with climate-hostile alternatives at moderate carbon prices. Without sharp increases in fossil fuel prices and without the results of cumulative R&D efforts in the coming decade, carbon prices are unlikely to rise globally to a level able to expand the share of renewables, clean fossil fuels or nuclear dramatically. This is also the phase in which Europe should embark on a shaping-oriented policy path to income generation for the poor, gearing energy R&D towards options that potentially diversify energy imports away from fossil fuel exporting, high-income countries towards renewables exporting, low-income countries. In the last and third phase, renewable resources have conquered an impressive part of world energy markets. The economic opportunities offered by exports of renewables will at the same time address global equity problems and problems of energy access for the poor. Just as fossil fuel price increases will proof to be a prerequisite for effective climate change policies in the second phase, so will global carbon price increases become a prerequisite for reducing global inequity in the third phase. This conclusion is based on the notion that the physical distribution of solar blessed territory and potentially arable land will prove to be a much more potent driver for improving global equity than the presently prominent tendency to rely on ethically justified distribution schemes for global carbon emission allowances. Fossil fuels may be scarce, but renewable resources are sparse. If owners of fossil fuels are rewarded the consequences are likely to reinforce global inequities, but if owners of renewable resources are rewarded the consequences are likely to promote global equity. Unfortunately, the impact of these physical factors on distributional issues and poverty reduction will only become prominent when both fossil fuel prices and carbon prices are very high.

[4]Even the International Energy Agency has left its previous, reassuring stance. Its latest World Energy Outlook carries a truly alarming message. See (IEA 2008).

2.3 *The Sunny Side of Doom*

One fundamental message of the vision packaged in this storyline is, that there is a sunny side to doom. The storyline suggests that Europe should guard against consecutive instances of doom first, but at the same time it also suggests that the results of the required hedging policies will help to solve the next problem, if adequate shaping policies for the next problem are tackled simultaneously. Europe will not be able to solve global sustainability problems unilaterally and it should prepare for unpleasant surprises in timely fashion. But at the same time, Europe can actively try to shape the direction of potential solutions for one problem in a way that may help to soften the blow of the next problem effectively. In fact, if such secondary shaping policies are well orchestrated with the primary hedging policies, Europe will be able to promote its own economic, environmental and social objectives much more effectively in later stages. The fact that Europe has already a policy in place for climate change mitigation is an important shaping oriented institutional innovation. But this important achievement will only be globally effective in the long run, when fossil fuel prices are much higher and climate change starts to have major economic impacts. By that time Europe will be in the best position to reap the benefits of the technological innovations that were driven by energy security concerns, but can now be geared towards solving climate change problems. Although environmental doom because of climate change is already upon us, it has yet to affect economic objectives in a major way all around the globe, before the threat will lead to globally effective policies on carbon pricing. But once those climate change policies are globally in place, they will have a major impact on equity related problems of energy. Existing global political priorities and existing global governance capacities may rank from economic via environmental to social issues, but ultimately each consecutive step in solving economic, environmental and social problems in the domain of energy and sustainability will inexorably bring the solution of remaining problems within reach. The sunny side of doom implies that sustainable ambitions in the sky should reflect political realities on the ground. When dealing with doom, it becomes politically much more acceptable to invoke the visible fist of government in addition to the invisible hand of market. This is what matters for painful policy decisions affecting the market place and the position of incumbent regime actors.

3 Illustration for the Transition to Sustainable Mobility

3.1 *Some Basic Facts About Prices and Costs in the Transportation Sector*

To illustrate the nature of effective policy design for the case of sustainable mobility let me present some basic facts about prices and costs in the transportation sector (see Fig. 2). Around June 2008 the price of gasoline was roughly 1.60 € per litre,

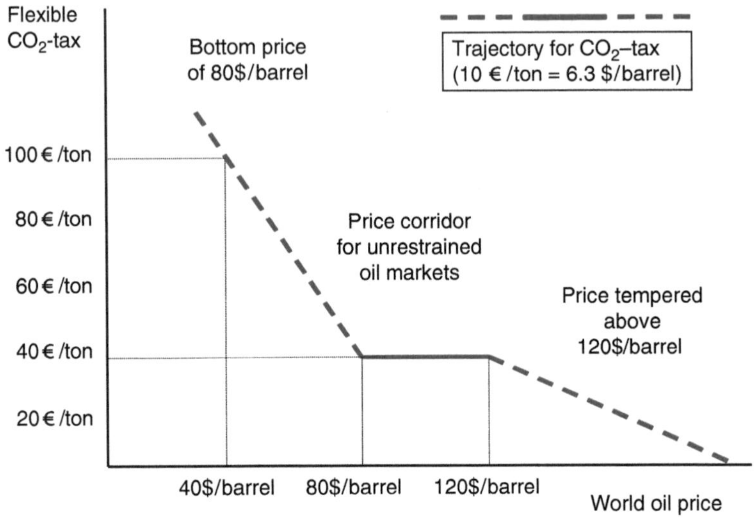

Fig. 2 Synchronizing world oil prices and CO_2-taxes

of which 19% value added tax, 69.5 cents excise duty, 13.5 cents distribution margin, and 57 cents gasoline product price.[5] This high price level had little effect on actual demand for gasoline, but it spread panic around Europe leading to orchestrated protests and political pressure to reduce taxes. This raises the question how badly permanent oil price increases would really hurt consumers in the long run, say by 2020. To answer this question meaningfully, one should realize that the average gasoline car is likely to be at least 20% more efficient by 2020. Similarly, we can assume that the average income of consumers will increase by some 2% annually in the mean time. Given unchanged fiscal policies and operating margins, this would allow a further increase of the oil price from 135$/barrel in June 2008 to over 300$/barrel in June 2020 without raising the budget share of gasoline costs in the average income of consumers (assuming dollar exchange rates constant at 1.5 $/€). This simple calculation serves to demonstrate, that even dramatic price increases of crude oil prices are unlikely to affect energy demand in a major way in the long run. But now consider the same question from the point of view of global fuel producers. The change from 135$/barrel to 300$/barrel would make the most exotic alternatives for conventional oil a valid economic proposition in the long run. In fact, most energy experts would consider a guaranteed price above 100$/barrel an appropriate level to introduce a wide range of conventional oil alternatives

[5]The figures are based on the actual price composition of gasoline in the Netherlands, since averages for Europe are not easily calculated. However, these figures represent the European situation sufficiently well for this illustrative case.

including unconventional oil (tar sands, heavy oil), gaseous fuels (GTL and CNG), biofuels (although probably not in a sustainable way from the point of view of land use and biodiversity), coal liquids, electricity and hydrogen (although probably not in combination with carbon capture and storage). High fossil fuel prices would thus be able to solve energy security problems, but they will not be able to solve climate change problems. First, they will have little effect on the volume of kilometres driven and sooner or later this will have a severe impact on the global potential to supply fuels sustainably. Secondly, they may lead to both lower and higher CO_2-emissions per kilometre.

3.2 *The Transition Tandem of Oil Prices and Carbon Taxes*

The end of cheap conventional oil will not automatically result in the end of climate-hostile transportation fuels. To the contrary, it may easily result in sharp increases of CO_2-emissions when economies feel forced to introduce worse alternatives such as unconventional oil and coal liquids. Of course, climate change policy makers will argue that CO_2-taxes are necessary to make competitive life for such alternatives difficult. Unfortunately, the potential level of such taxes in the period up to 2020 is by far insufficiently high to make much of an impact. The carbon content of a barrel of crude oil is 418 kilo CO_2. A carbon price of 10 €/t CO_2 is thus equivalent to a price increase of roughly 4.2 €/barrel or 6.3 $/barrel. Expectations about carbon price developments in the European carbon market in the period up to 2020 generally range from 20 to 60 €/t CO_2. This is equivalent to an oil price increase between 12 and 40 $/barrel. In the half year since June 2008 oil prices have tumbled from 135 €/barrel to 50$/barrel. If CO_2-taxes would have been raised to prevent such a tumble in order to keep climate-friendly alternatives alive, they would now have to amount to 135 €/t.[6] In other words, developments on the crude oil world market in the last 6 months of 2009 have had a negative impact on the competitive position of climate-friendly fuels that is almost four times as high as the impact of developments in CO_2 prices that the climate policy community generally expects for the 12 year period up to 2020. This simple calculation does not serve to demonstrate, that CO_2-pricing is useless. It only demonstrates that CO_2-pricing might become useful once world oil prices are very high, because at that level they are able to bridge the gap between climate-friendly and climate-hostile alternatives. In other words, we must wait for all alternatives, clean and dirty, to become attractive, and than carbon taxes can work in tandem with oil prices to tilt the balance towards the climate-friendly alternatives.

---

[6]The number of 135 €/t for the required CO_2-tax level happens to be equal to the number of 135$/barrel for the price of oil in June 2008. This is purely accidental.

3.3 The Transition Tandem of Fuel Prices and Road Charges

Even if we succeed in bringing sustainable fuels to the market successfully once both world oil prices and CO_2-taxes are very high, we are still stuck with the fact that high fuel prices have comparatively little impact on tempering mobility volume. Ultimately, it will become progressively more difficult to produce the required fuels in a sustainable way when global incomes continue to grow and converge. For instance, if we rely in part on sustainable biofuels, this would undoubtedly lead to replacing the present peak oil problem with a future peak soil problem. This is the stage where an increasing reliance on road charges is of eminent importance and where policy innovations are needed that are able to change people's expectations about mobility dramatically. The problem with today's mobility policies is that they concentrate far too exclusively on technological solutions. In the long term much more reliance on life style changes is inevitable. Regime changes towards sustainable mobility must include both technological and behavioural components. Of course, the discourse about such measures is politically unpalatable at the moment and such a discourse requires a far greater emphasis on equity related distribution issues than efficiency related technological issues. Just as oil prices and CO_2-taxes should form a synchronized tandem on the way to sustainable mobility in the first transition phase, so should fuel prices and road charges form a synchronized tandem on the way to sustainable mobility in the second transition phase.

3.4 Hedging Policies for Managing Sustainable Mobility Transitions

Sustainable mobility in the long run is best served by starting out with a hedging strategy for energy security coupled with a shaping strategy for climate change. The first element of such a hedging strategy would involve the early creation of a viable technological alternative for the liquid fuel, internal combustion car in the form of the electric plug-in hybrid. This is equivalent to chosen a winning technology that leaves open a host of options with respect to the ultimate source of electricity in the long run. Moreover, liquid fuels will serve as range extender, either from sustainable biofuels or from clean fossil fuels. Ultimately, hydrogen would still remain an option. Its future should be viewed however as a potential option at the far end of the electric plug-in hybrid development trajectory rather than as a competing trajectory at the start of sustainable mobility transitions. Electrical alternatives are much less dependent on time consuming, large-scale system innovations than hydrogen alternatives and are benefiting from parallel innovation trajectories elsewhere in a major way (for instance battery development for mobile ICT). The second element of a hedging strategy would involve establishing a CO_2-tax in the transport sector. This CO_2-tax would not be used as major instrument to implement climate change policies at this stage nor should it raise prices in the initial stages of implementation. It would simply replace part of existing national excise taxes and it would be used for two purposes: managing world oil market prices to enhance energy security and avoid economic disruption and funding European-wide R&D budgets for climate change mitigation and adaptation. The synchronization of world oil market prices and CO_2-taxes could be implemented by establishing a flexible tax that moves in the opposite direction as world oil prices (see Fig. 2). When world oil prices move below a fixed bottom price CO_2-taxes would be raised to guarantee the market for alternative producers of fuels. When world oil prices rise above a fixed ceiling price CO_2-prices would be lowered gradually to ease economic disruption. The corridor for a stable CO_2-tax would reflect expected CO_2-prices in the cap-and-trade market.

3.5 Shaping Strategies for Managing Sustainable Mobility Transitions

Choosing the electric plug-in hybrid as a winner on the European level, may be a good hedging policy for the first phase of a European energy strategy, but sooner or later this will simply shift the burden of sustainable energy to the electricity sector. Moreover, the need for finding a sustainable, liquid range extender will also remain. And if we extend our view to the final phase of the sustainable mobility transition and problems of global inequity must also be addressed, it is not a wise strategy to replace global peak oil problems with global peak soil problems. A complementary, shaping strategy for managing sustainable mobility would thus contain two major elements of climate change policies. The first element involves expanding our R&D efforts to provide sustainable electricity and biofuels in an affordable and reliable way. This involves a much more determinate effort to make offshore wind a success and an intensified effort to explore the potential for affordable solar electricity including concentrating solar power. Europe should also explore the feasibility of sustainable biofuel importation at an early stage. The latter two options should proceed through well-funded technological development agreements between Europe and developing nations. The second element involves policy innovation, not technology innovation. It is unlikely that any sustainable technology, even in the far future would allow continued expansion of kilometres driven. We should therefore explore much more intensively how to device fiscal policies to deter mobility wisely. This is as much a question of equity as of efficiency and requires a complete revision of the fiscal system including measures such as progressive road charges in return for flatter income taxes.

4 Implications for European Energy R&D Ambitions

4.1 European Energy R&D Governance Needs Strengthening

European climate change policies have undoubtedly had an enormous catalytic effect on the implementation of renewable energy and energy efficiency measures in member states. Moreover, regardless of its limited impact so far in terms of actual investments, the establishment of a cap-and-trade system for CO_2-emissions from large installations is a major policy innovation that has created a solid market pull mechanism for energy technology innovation. It has also made Europe an uncontested frontrunner in the global governance of climate change problems. With respect to the technology push part of the innovation cycle, Europe's position may be less prominent. The average performance of Europe in innovation indicators compared to the US and Japan remains mediocre.[7] Of course, steps have been taken to redress the balance. Energy and climate form an important theme when it comes to implementing the Lisbon objectives related to economic growth, social welfare and environmental quality. Nevertheless, European energy innovation instruments are generally viewed as relatively ineffective in promoting innovation. Integration of the three corners of the knowledge triangle (education, research and innovation) remains problematic. National priorities in R&D are dominant and European cooperation is fragmented. Finally, the level of energy R&D resources (both human and financial) is dismally low in view of the three energy-related faces of doom. In the first part of this paper, I have argued that in the case of climate change Europe should adopt a shaping strategy rather than a hedging strategy and that such a strategy implies a shift away from implementation policies to R&D policies in the climate change domain. How this could happen will be an important challenge for dealing with doom.

4.2 Financing Is Key Challenge for Improved European R&D Governance

To speed up innovation Europe has embarked on renewing its energy R&D instruments. The two new instruments are the Strategic Energy Technology plan (SET-plan) and the European Institute of Innovation and Technology (EIT). The SET-plan proposes to establish a European Energy Research Alliance (EERA). Unfortunately, there is no additional budget for energy R&D on the European level available to finance the actual research that these institutions are meant to promote and coordinate. Without additional budgets these institutions are likely to cannibalize existing funds both at the European level (from FP7 onwards) and at the national

[7] For a recent performance overview see (EC 2008).

level (in member states keen to join the European bandwagon). Additional funds could come from either private or public resources. The chances of such funds coming forward any time soon seem remote. Policy makers publicly state that private funds are absolutely necessary to guarantee that the famous valley of death between R&D and commercial applications is automatically closed. Company decision makers privately state that they are already paying taxes and that it is up to governments to fund pre-competitive research and set public priorities right. To complicate matters, it is also generally true that private R&D budgets flow unconstrained across European borders, but that national taxpayers are notoriously wary of seeing their tax contributions disappear across the border. To provide future funding of R&D budgets at the European level, the biggest challenge would be to design European energy policies in such a way, that they automatically generate additional funds for European wide R&D priorities. Linking hedging-oriented implementation policies for energy security with shaping-oriented R&D polices financially may turn out to be the most promising possibility. In the case of sustainable mobility policies a CO_2-tax designed to work in tandem with world oil price developments could be partially earmarked for R&D purposes in the domain of sustainable electricity and biofuels. Such earmarked funds should include a generous budget for financing technological development agreements with developing nations. In general, policy makers tend to be averse to hypothecated taxes, because it ties their hands in the sense of leaving less room for discretionary spending of revenues. But in the case of the combined challenges of energy scarcity, climate change and global inequity as discussed in this analysis, this may be precisely what is needed for long-term success.

References

EC. (2008, February). *European innovation scoreboard: Comparative analysis of innovation performance*. Brussels: European Commission Directorate-General for Enterprise.

Frei, C. W. (2004). The Kyoto protocol – a victim of supply security? Or: if Maslow were in energy politics. *Energy Policy, 32*, 1253–1256.

IEA. (2008, November). *World Energy Outlook 2008*. Paris: International Energy Agency.

NIC. (2008, November). *Global trends 2025: A transformed world*. Washington, DC: US National Intelligence Council.

Silberglitt, R., Hove, A., & Shulman, P. (2003). Analysis of US energy scenarios: Meta-scenarios, pathways and policy implications. *Technological Forecasting and Social Change, 70*, 297–315.

A Transition Research Perspective on Governance for Sustainability

Derk Loorbach, Niki Frantzeskaki, and Wil Thissen

In this chapter we present the transitions approach as an integrated perspective to understand and possibly orient our society towards sustainable development. Since the concept of sustainability is inherently normative, subjective and ambiguous, we argue that (unlike some more traditional approaches to sustainable development) we should focus on an open facilitation and stimulation of social processes towards sustainability. The transitions approach and transition management specifically, seek to deal with ongoing changes in society in an evolutionary manner so as to influence these ongoing changes in terms of speed and direction: towards sustainability. A transitions approach to explore sustainability transitions poses novel challenges for research: there are no unequivocal answers, nor it is clear how these processes should be governed. We conclude our analysis by formulating the basic research questions central to the search for governance for sustainability, and by reflecting on the role of science in sustainability transitions.

D. Loorbach (✉)
DRIFT – Dutch Research Institute for Transitions, Faculty of Social Sciences, Erasmus University Rotterdam, Burgemeester Oudlaan 50, P.O. Box 1738, 3000 DR Rotterdam, The Netherlands
e-mail: loorbach@fsw.eur.nl

N. Frantzeskaki
DRIFT – Dutch Research Institute for Transitions, Faculty of Social Sciences, Erasmus University Rotterdam, Burgemeester Oudlaan 50, P.O. Box 1738, 3000 DR Rotterdam, The Netherlands

Faculty of Technology, Policy and Management, Delft University of Technology, Jaffalaan 5, 2628BX Delft, The Netherlands
e-mail: frantzeskaki@fsw.eur.nl

W. Thissen
Faculty of Technology, Policy and Management, Delft University of Technology, Jaffalaan 5, 2628BX Delft, The Netherlands
e-mail: W.A.H.Thissen@tudelft.nl

C.C. Jaeger et al. (eds.), *European Research on Sustainable Development*,
DOI 10.1007/978-3-642-19202-9_7, © European Union, 2011
Published by Springer-Verlag Berlin Heidelberg 2011

1 Introduction

Over the last decade, sustainable development has become a central concept guiding scientific debates and policies related to complex and persistent problems (Jansen 2003; Meadowcroft 2000; Scott and Gough 2004). Sustainable development aims to ensure economic welfare, social equality and ecologic quality across society, generations and into the future. It is commonly associated with those types of social problems that demand a fundamental restructuring of dominant paradigms, institutions and practices. In such a context, conventional forms of planning and policy as well as traditional conceptions of knowledge and of the role of scientists are being challenged (Flyvbjerg 2001; Kates et al. 2001). Over the last decade, new research and policy fields have emerged adopting more integrated perspectives and concepts, such as transition and resilience approaches that are explicitly based on complexity and uncertainty.

In the transition perspective, our society is composed of complex adaptive systems in which individuals and organizations (may) self-organize within the limits set by physical, institutional and informal structures, and (can) experience the emergence of innovations of different types (e.g. technological innovations or social innovations in the form of new practices and/or routines). Historical transitions such as energy supply or mobility transitions were semi-autonomous societal processes; since the vast majority was not steered in a collective way but emerged as a societal outcome. Transitions can take decades to materialize and are highly uncertain in terms of future development, possibilities for change and the level of intervention possible in such dynamic processes. Though it is clear that ongoing processes of change need to be oriented towards more sustainable system's states, the often blurred vision of what exactly is sustainable makes the pathways towards a sustainable system state highly uncertain.

For facing the challenge of sustainability, not only the translation of sustainability in a specific context needs to be coordinated and to comply with the sustainability values, but also the actors involved need to re-evaluate their roles and practices. We argue in this chapter that the transitions approach can aid our understanding of sustainability transitions as well as our drawing of governance guidelines[1] that encounter the multi-faceted nature of sustainability and the complex nature of societal transitions.

[1]We perceive governance as a meta-level pattern of societal interactions (intended and unintended) or as Kooiman (1993, p. 2) notes, governance to be interactive as "the pattern that emerges from the governing activities of social, political and administrative actors" that "focuses on the interactions taking place between governing actors within social-political situations." (Kooiman 2003, p. 7).

In our effort to present what the transitions approach can offer to sustainable development research, we will first elaborate on existing propositions of sustainability and sustainable development (Sect. 2). In our analysis of sustainability research and the different research streams and approaches that are risen by sustainability research, we will present (how and) that the transitions approach also relates to sustainability research (Sect. 3). The transitions' approach takes into account the intrinsic characteristics of the societal system (e.g. the complexity, the interdependence of system components, and the presence of multiple actors and networks) and perceives fundamental changes in culture, structures and practices as an answer to persistent problems (Frantzeskaki and De Haan 2009). Based on the transitions approach, we will draw governance guidelines for sustainability transitions that also relate to and give rise to research propositions for governance of transitions (Sect. 4). In addition to this, the role of science and scientists for sustainability transitions will be discussed and a reflection on the new roles of scientists will be given (Sect. 5). Concluding remarks regarding the analysis will be presented at the end (Sect. 6).

2 Sustainability and Sustainable Development: Existing Propositions

Since the late 1980s, many countries have committed themselves to sustainable development and are struggling with how to achieve it. Following the Brundtland report *Our Common Future* (WCED 1987), sustainable development came to be defined as redirection of social development in ways that combine prosperity, environmental protection and social cohesion. In the Brundtland report, sustainable development was defined as a development that meets the needs of the present generation, without compromising the needs of future generations (WCED 1987). This definition is normative since future generations should have the same possibilities, subjective since it requires an assessment of what these future needs are, and ambiguous since these future needs are determined by cultural, ecological and economic developments that can be defined in more than one way (Martens and Rotmans 2002; UN 1997).

At the international level, there is a consensus on the need for sustainable development and key areas in which the next decade significant progress needs to be made: poverty, hunger, health, education, life expectancy, environmental sustainability and global partnerships (UN 2005). The approach to sustainable development adopted by the United Nations is to realize overall consensus while allowing for a variation of strategies and solutions to be chosen by individual countries, regions and actors at different levels (UN 2005). This means that in practice, different countries have taken up different strategies to cope with the challenge of achieving sustainable development. A lot of countries opted for sustainability councils and the development of sustainability indicators (see Mulder 2006, pp. 148–165). In this context, sustainable development has been represented as the intersection of economic, social and environmental agendas and the need to integrate (predominantly) environmental concerns into regular policies.

We can derive some basic characteristics that are attributed to the concept of sustainable development that are visited in almost all definitions and scientific writings. The first is that sustainability is intergenerational. This means that a long-time horizon, at least one or two generations (25–50 years), has to be considered. The second characteristic is the importance of scale. Sustainability can be achieved at different levels. Pursuing sustainability at one level does not necessarily scale-up or scale-down to different levels. Hence, local or regional sustainability does not necessarily mean national or global sustainability and vice versa. Consequently, sustainability analysis requires a multitude of scale levels. The third common characteristic is that sustainability relates to multiple domains. Sustainability encompasses a certain context-specific balance between ecological, economic and socio-cultural values and stakes (Kates et al. 2001; Pezzoli 1997). In addition to this, sustainability values can form and be adopted by different sectors such as energy, food or water.

Sustainable development is therefore a normative orientation that provides a frame of reference to discuss and direct differences in perception, ambition and understanding between actors in light of desired changes in society. After the initial optimism during the 1990's about win-win opportunities, it is increasingly understood that there are tradeoffs between different values and interests in any type of development (at least in the short term) and that each development tosses up new problems for society. Considering the presence and impact of the aforementioned impacts and trade-offs, it is argued that alternatives for development can only be called sustainable when they are (co-)developed, implemented and formulated by societal actors (Clark 2003).

Following this, we suggest that sustainable development should be considered as a continuous process in which societal values and interests are represented, negotiated and balanced. At the same time, new alternatives and visions need to be explored and experimented with. Sustainable development is a multi-dimensional, dynamic and plural concept that neither can be translated into the narrow terms of static optimization nor is conducive to strategies based on direct control. This is a distinctive characteristic of sustainable development as a new type of development process vis-à-vis economic development: the goal of sustainability exists but its target level changes over time due to its redefinition by every generation (Mulder 2006, p. 74). As Meadowcroft (1997, p. 37) phrases this perspective: "Each generation must take up the challenge anew, determining in what directions their development objectives lie, what constitutes the boundaries of the environmentally possible and the environmentally desirable, and what their understanding of the requirements of social justice is".

Arguably, sustainable development as a broad notion of an integrative and balanced, yet flexible societal development should be used as guiding principle for future-oriented actions. This means that the challenge of sustainable development can be formulated as a continuous governance process that enables representation of various perspectives, values and interest and creates space for experimentation, innovation and learning.

3 Transitions Approach for Sustainability

The focus of the current analysis is on *sustainability transitions* or transitions to sustainability that concern continuous processes of change that reorient and restructure a societal system towards a sustainable system state that satisfies sustainability values. In line with this and as already indicated in the introduction, adopting a view on the transition to sustainability implies an integrative view of sustainability, which is capable of incorporating multiple domains, multiple levels of scale and spans a long-term (being intergenerational).

3.1 Foundations of Transitions Approach

In the early 1990s complex systems theory was introduced, focusing on the co-evolutionary development of systems. The establishment of the Santa Fé institute in New Mexico in the United States in 1984 functioned as incubator for a new research movement, which laid the basis for complex adaptive systems theory (Holland 1995; Kauffman 1995). Although the theory is far from mature, it has attracted a great deal of attention and has many applications in diverse research fields: in biology (Kauffman 1995), economics (Arthur et al. 1997), ecology (Gunderson and Holling 2002) and public administration (Teisman and Klijn 2008). The basic idea is that complex interactions between different elements can be understood in a systemic sense: through their interaction, elements within a system co-evolve with each-other and with their environment, new structures and novelties emerge and new configurations appear through self-organization.

The basic mechanisms that underlie change in complex adaptive systems are co-evolution, emergence and self-organization (Holland 1995). Societal systems can be considered as complex adaptive systems. Societal sectors consist of numerous interlinked elements (e.g. actors and institutions), there is a high degree of uncertainty about their interactions and feedback and they have an open and nested character in terms of different levels of organisation. From this perspective, typical complex system behaviour can be recognized, as for example emerging structures, co-evolving (policy) domains and self-organizing processes can be observed. One of the possible patterns distinguished is that of transition: a system in a relatively stable equilibrium is (suddenly) going into a phase of rapid change through a process in which self-organisation and co-evolution play an important role before a new equilibrium is found.

3.2 Transitions Perspective on Systems Innovation

History has witnessed numerous transitions in economy, agriculture, mobility, and energy, but also in areas such as education, health care, and social structure (Geels 2004; Rotmans et al. 2001). Transitions are processes of 'degradation' and 'break-down' as well as of 'build up' and 'innovation' (Gunderson and Holling 2002) or of 'creative destruction' (Schumpeter 1934) of societal systems. The central assumption is that societal systems go through long periods of relative stability and optimization that are followed by relatively short periods of radical/fundamental change. These changes can be analyzed in terms of multi-level (distinguishing between dominant regimes, upcoming innovations/niches and landscape development (Geels 2004)) and multi-phase frameworks (in terms of predevelopment, take-off, acceleration and stabilization) (Rotmans 1994).

3.3 Transitions Perspective on Systems' Sustainability

Historically, transitions have been primarily driven by changes in social subsystems that initiated large-scale changes such as demographic growth, technological innovation or economic expansion. In a sense, these historical transitions (such as those part of the industrialization era, the post-war emergence of mobility, intensive agriculture or fossil energy systems), were also partly driven by the promise of solving societal problems such as poverty, inequality, education and so on. Such transitions however produced, in dealing with certain issues, their own problems in turn. While individuals might now have availability of cheap energy and mobility, it has co-produced for example pollution, resources' exploitation and congestion. In that sense, the transitions leading to our current modern society have had as side-effect the current environmental problems. The challenge in dealing with modern complex and persistent problems is to find new ways in dealing with them in a more anticipatory and exploratory manner.

While complex processes of change are occurring, we need to try to better understand their dynamics and try to influence their pace and direction. Combined with the basic notion that sustainability is ambiguous, uncertain and contested, this means that the only way to 'enable' sustainable development is through participatory processes in which sustainability is discussed, negotiated and explored in light of the major changes that are undoubtedly necessary.

A process-philosophy for achieving sustainable development may aid the development of concrete action and allow for plurality in actor's objectives and actions as well as for flexibility of processes and actions. An example of such a participatory tool is the transition arena (Loorbach and Rotmans 2010) that aims at achieving fundamental change in practices and visions by involving and facilitating frontrunners. Within the transition arena, frontrunners were facilitated to work collectively to transform social systems towards sustainability in the long-term, as well as to compete over 'best' solutions and conflicting values on the short-term. The transition arena is a tool with its limitations. Based on the principle of small-group effectiveness, the transition arena as a participatory tool raises issues about democratic legitimacy, accountability and control (Shove and Walker 2007; Hendriks and Grin 2007). At the same time, the deliberate visioning process that takes place within the transition arena questions the legitimacy of the existing dominant institutions since they are not able to include uncertainty and/or to create the level of social innovation needed for transitions (Hendriks and Grin 2007).

Dominant (policy and research) approaches predominantly seek to improve existing systems, leading to gradual improvement. Opposing this, transitions' thinking suggests that a sustainable development process requires a fundamental shift of a societal system. Instead of incremental changes that aim at preserving existing functioning, transition thinking focuses on radical changes or "transformation of both (...) systems and social structures and practices" (after Meadowcroft 1997, p. 430). It is concluded that apart from a co-evolving target of sustainability that a society needs to form, transition guidelines[2] are essential in creating space for and enabling a societal transition to sustainability.

Methodologically, the new research field of transitions requires new types of research that have an integrative nature, are normative in their ambitions, have a desire to contribute to societal change and are participatory. Over the last decade, a number of such new types of research have emerged. Examples include the Integrated Assessment, Post Normal Science and Action Research. Although these examples are partly grounded in and based on existing approaches and methods, they all provide a new way of formulating and directing the research process and the role of researchers. They therefore provide a valuable basis for conceptualizing 'transition research'. Transition management is an example of a research topic that by definition cannot be developed in a traditional, purely scientific sense. It is based on transitions approach, itself still a theory (or approach) in development, and presumes that an understanding of transitions can lead to other types of (policy) practices.

The only way to achieve coherence between theory and practice of transition management is through a learning-by-doing and doing-by-learning approach in which fundamental research, theory development, participatory research and applied research are combined (Loorbach 2007). The research methodology is unfolding during the research process: as new theoretical insights emerge, experimental and exploratory cases are used, and vice versa. When observations about operational processes inform or challenge theory they need to be structured, integrated and grounded.

[2]Transition governance guidelines are process-oriented propositions that relate to the process design towards fundamental change or transformation. Transition governance guidelines are not concerned with the definition of targets, or goals but with the design or the framing of the actions that will take place over the course of a system transition.

3.4 Integrated Assessment, Post-normal Science and Sustainability Science

The transition concept is an Integrated Assessment concept. Integrated Assessment (IA) (Rotmans 1998) is defined as a scientific "meta-discipline" that integrates knowledge about a problem domain and makes it available for societal learning and decision making processes. It is a relatively recent field that emerged during the 1990s and is explicitly concerned with providing policy-relevant knowledge for complex societal problems. Because of this ambition and the focus on complex problems, Integrated Assessment by definition is both interdisciplinary (integrating scientific disciplines) and transdisciplinary (integrating scientific and lay knowledge). Integrated Assessment has been used as a new assessment paradigm in, for example, modelling, scenario-based research, and complexity research, but has also influenced the debate on the role of research for policy making. Integrated Assessment has a history of being applied first and foremost in the area of long-range and long-term environmental policy issues, but has developed to a research approach accepted in the policy arena to be supportive for long-term policy planning processes. By definition, transitions cannot be understood from a single scientific discipline or societal perspective and thus require interdisciplinary tools and frameworks. Transitions approach and transition management have primarily been based on knowledge and insights from Integrated Assessment.

This consideration also relates to the concept of Post-Normal Science (Funtowicz and Ravetz 1994; Ravetz 1999) that legitimates the involvement of diverse knowledge sources in science for policy through calling for extended peer communities and emphasizing the inherent uncertainties and values in policy-related science. A key notion in Integrated Assessment and in Post-Normal Science is the acceptance of uncertainty and ambiguity, which necessitates a participatory research approach or at least a structured form of interaction between researchers and societal actors to produce policy-relevant knowledge. Sustainable development in this context is, according to the field of Integrated Assessment, a possible normative orientation that provides a frame of reference to discuss and direct differences in perception, ambition and understanding between actors. The rationale behind this is that solutions for sustainable development can only be called sustainable when they are (co-)developed, implemented and sustained by societal actors (Clark 2003). This means that scientific knowledge related to sustainable development is not a goal in itself, but rather a means to achieve progress. From this perspective, a modest and vulnerable position of a scientist in the process of sustainable development is required, rather than the position of provider of objective truths or that of outside reflector producing policy-advice as an end-product of research. The objective position of research(ers) related to policy, and in general the science-policy interface has already been the subject of debate for decades (Wildavsky 1979; Hisschemöller and Hoppe 1996), but has been revived in the context of sustainable development, where scientific knowledge as well as political and social knowledge are all as ambiguous as the solutions and outcomes (Hisschemoller et al. 2001).

A field closely related to Integrated Assessment and transition research is that of Sustainability Science (Kates et al. 2001; Kasemir et al. 2003; Clark et al. 2005). Stemming from the field of science and technology, 'sustainability science' has emerged as a (somewhat controversial) term depicting those developments within scientific disciplines that deal with sustainability issues, increasingly in cooperation with practitioners. Without being as defined and concrete as Integrated Assessment, Sustainability Science is more or less a general term for a development in science as a whole towards more multi- and interdisciplinary research related to complex societal issues. Sustainability Science mainly refers to the field of global environmental and sustainability research and emphasizes the importance of the involvement of stakeholders in the knowledge development process. While Integrated Assessment offers concrete tools and methods for complexity and sustainability research, Sustainability Science redefines the role of research and researchers at an abstract level. For transition research this is relevant, since the ambitions behind transition research are similar to those behind Sustainability Science: scientific and societal impact based on an active and participatory role of researchers.

The participatory nature of Sustainability Science has been theorized and methodologically underpinned over the last decade. In fact, participatory knowledge development, aimed at integrating practical/tacit and scientific knowledge, has become a new field of research in itself (Van Asselt and Rijkens-Klomp 2002; Kasemir et al. 2003). The central issue in this field is that participation in practice is often unstructured and ad hoc. Consequently, methods and tools for both participatory policy making and participatory (integrated) research need to be developed and tested (Van de Kerkhof 2006). Although participatory methods (e.g. focus groups, consensus conferences, scenario exercises, simulation games) have a long history, they have been reinterpreted in the context of Integrated Assessment with regard to the profile of the participants, the goal of participation and the degree of participation (Rotmans 1998). So far, participation has mainly been used in the context of policy-making (to generate public support) and has been underdeveloped in scientific research as a means to generate knowledge with a higher relevance for society. The rationale behind participation in research such as Integrated Assessment and Sustainability Science is that the knowledge generated is not only relevant for the situation it is developed for, but that the participants have already during the knowledge development process internalized some of the knowledge generated, which enhances the chances for application of the knowledge. The participatory approach behind transition research serves these two main goals: development of new knowledge and also application of this new knowledge and through that change in real-life. The participatory approach is thus an instrument for the transition researcher to transfer knowledge as well as to develop new theory.

4 Governance for Sustainability Transitions

Sustainable development as a societal objective is a continuous intergenerational, multi-scale (global and local) and multi-domain process of seeking a balance between social, economic, ecological and cultural values. Following the premises underlying transition research, transitions themselves are, like the concept of sustainable development, inherently uncertain, complex and unstructured. They cannot be studied in a classical straightforward manner, nor can they be influenced through linear blueprint approaches. The role of transition research in understanding and shaping ongoing transitions therefore necessary has characteristics of social construction, a concept well known in technology studies (Bijker et al. 1987). Applied to societal complexity, social construction means collectively structuring, identifying and giving meaning to ongoing changes as transitions. Transition management builds upon this idea to organize collective anticipation of future dynamics and structuring activities related to this, such as envisioning (or vision building processes), scenario-building, experimenting and monitoring. Transition management thus seeks to 'construct' a narrative of the need and possibility of a sustainability transitions, that will only be achieved if actors in society themselves make it 'true'.

Societal contexts are always unique, hence transition management cannot offer blueprints about governance means that can be effective in all contexts. Transition management can offer basic governance guidelines that can be used depending on the context. In the following section we start from four basic propositions to illustrate the view of transition approach for sustainability and how transition management as a governance approach can aid the achievement of sustainability with transition governance guidelines. For every governance principle discussed above, a number of governance guidelines are presented that are resulted by the operationalization of the governance principle into guidelines in compliance with the transition management principles (Loorbach 2007). Given that the transition governance guidelines offer only a direction towards action and are neither "blueprints" nor "concrete prescriptions for action", issues that require further research for achieving sustainability transitions are linked to the four sustainability governance principles.

4.1 Transition Guidelines and Research Questions

Research on transition dynamics and transition management resulted in a number of starting transition governance guidelines that are related to the below presented propositions. Given the complexity of phenomena as societal transitions, the below listed process guidelines are not set in stone but can and will evolve due to scientific debate and practical implementation. This approach is fundamentally different from a (more) descriptive and analytical scientific approach that would primarily focus on understanding these processes and describing them. The questions formulated below are in line with this based upon recent theoretical debates and empirical insights around transition management. An additional remark is that future research on sustainable development will require a focus on new modes of governance for promoting sustainability transitions.

Proposition 1: *Sustainability transitions are long-term processes of fundamental change that incorporate processes of societal, ecological, economic, cultural and technological evolution*

Transition governance guidelines:

(a) *Explore and understand the dynamics of the system so as to create feasible means for governance.* This implies that substance and process are inseparable. Process management on its own is not sufficient – insight into how the system works is an essential precondition for effective management. Systems-thinking (in terms of more than one domain (multi-domain) and different actors (multi-actor) at different scale levels (multi-level); analyzing how developments in one domain or level interact with developments in other domains or levels) is necessary to be able to take into account such possible means and leavers for intervention.

(b) *Enable learning-by-doing and doing-by-learning.* Social learning is a pivotal aspect of societal transition processes, aimed at 'reframing', changing the perspective of actors involved. Social learning as the combined outcome of learning-by-doing and doing-by-learning actions stimulates the development of visions, pathways and experiments that form a new context as well as pave the ground for the reorientation of the societal system.

Research questions:

- Which are the prevailing patterns of societal transitions in the form of multi-domain processes of change?
- Can we distinguish different types of transitions related to sustainability issues and what does that mean in terms of societal dynamics?
- Is it possible to understand ongoing transitions in which we all are part and if so, are we able to influence these?

Proposition 2: *Enabling societal processes of change (transitions) implies an integrated understanding of the dynamics of change and deliberate and reflexive strategies so as to allow for self-orientation of society towards a sustainable development pathway*

Transition governance guidelines:

(a) *Enable radical change in incremental steps.* Radical, structural change is needed to erode the existing structure of a system and ultimately dismantle it. Immediate radical change, however, would lead to maximal resistance from the deep structure, that cannot adjust to a too fast, radical change. Abrupt forcing of the system would disrupt the system and would create a backlash in the system because of its resilience. Incremental change allows the system to adjust to the new circumstances and to build up new structures that align to the new configuration.

(b) *Objectives should be flexible and adjustable at the system level.* The complexity of the system is at odds with the formulation of specific objectives. With flexible evolving objectives one is in a better position to react to changes from inside and outside the system. While being directed the structure and order of the system are also changing, and so the objectives set should change too.

Research questions:

- How can we better understand complex evolutions in society to make better use of complex system dynamics?
- Which are the means for governance that can deliberately promote societal transitions while allowing self-organization and self-orientation of the societal system?

Proposition 3: *Innovation and sustainable development are interlinked*

Transition governance guidelines:

(a) *Create space for niches in transition arenas and transition experiments.* A niche is a new structure, a small core of agents that emerges within the system and is seen as the incumbent for innovation. An emergent structure is formed around niches to stimulate the further development of these niches into (policy) regimes.
(b) *Focus on frontrunners.* In this context we define frontrunners as agents with peculiar competencies and qualities: creative minds, strategists and visionaries. Frontrunners are able to accelerate and/or to initiate the dissipation of structures in complex systems and are active at different levels and domains.
(c) *Guided variation and selection.* Diversity is required to avoid rigidity within the system. Rigidity here means reduced diversity due to selection mechanisms which means that the system cannot respond flexibly to changes in its environment. Rather than selecting innovative options in a too early stage options are kept open in order to learn about the pros and cons of available options before making a selection. Collective choices are made "along the way" on the basis of learning experiences at different levels.

Research questions:

- Which modes of governance can promote innovation while securing sustainability values in multiple domains?
- Which are means for governance that can create space for innovation that complies with sustainability values apart from regulation and institutionalization of innovation systems?
- Which are means for governance that can anticipate multiple visions and values of sustainability while facilitating the adoption of innovation?

Proposition 4: *Sustainability transition is a continuous open-ended process of societal innovation. Governance for sustainability transitions has thus to secure sustainability values such as long-term orientation and intergenerational justice*

Transition governance guidelines:

(a) *Long-term thinking as a framework for shaping short-term policy in the context of persistent societal problems.* Processes of back- and fore-casting: the setting of short-term goals based on long-term goals and the reflection on future developments through the use of scenarios.
(b) *Anticipation and adaptation.* Anticipating future trends and developments, taking account of weak signals and seeds of change acting as the harbingers of the future, is a key element of a pro-active, long-term strategy as transition management. This future orientation is accompanied by a strategy of adaptation, which means adjusting while the structure of the system is changing.

Research questions:

- How can governance deal with the tension between promoting continuous innovation while at the same time needing to ensure institutional performance?
- Which are the means for governance that incorporate long-term orientation and its uncertainties?
- Which are the means for governance that ensure reflexivity and adaptability in face of long-term processes of transitions to sustainability?

5 Role of Science in Transitions

Transition research combines traditional and new types of research and the underlying methodological approach is thus based on integration and combination of methods linked to the specific research context and questions. The impact of the concept of transition management on policy-making and the rapid development within the scientific community shows a clear need for this type of participatory, normative and integrative research next to more regular scientific research and knowledge. The major weakness of the new type of research sketched here is that much depends on the abilities of the researchers and practitioners involved to communicate so that they understand each other, can co-produce knowledge that is scientifically and socially relevant and develop solutions that are actually used in practice. Especially for researchers, the reflexive component that now becomes part of the research process (i.e. feeding back insights from practice into the theory) requires finding adequate ways of dealing with a diversity of participants, differing normative interests and ambitions, and a variety of sources of knowledge. Development of competences and skills that enable the researcher to fulfil the required roles linked to research ambitions through training and experience should thus also form an explicit part of a transition research process.

The only way to achieve coherence between theory and practice of transition management is through a learning-by-doing and doing-by-learning approach in which fundamental research, theory development, participatory research and applied research are combined. The research methodology is unfolding during the research process: as new theoretical insights emerge, experimental and exploratory cases are used, and vice versa when observations about operational processes inform or challenge theory they need to be structured, integrated and grounded.

Our working recommendations for the research for sustainability transitions concern the role of scientists in transforming policies and practices towards sustainability. Our experience in interdisciplinary and inter-sectoral projects showed us that research for sustainability requires not only an action research perspective but also simple actions that can result in desirable changes towards sustainability. Driven by our lessons, we recommend the following four roles for scientists that want to have an active role in sustainability transitions:

- *Create interdisciplinary teams for research projects*: Interdisciplinary research teams appear effective in linking knowledge from different fields and in this way connecting sustainability to different practices and applications. It may seem trivial to recommend interdisciplinarity, we know however that not all research programs are conducted by interdisciplinary teams within the sustainability field. The advantages of an interdisciplinary research team that includes policy scientists or management scientists for example are that the research findings can be translated into operational actions and can be connected to existing practices. Such operational outputs may result in achieving sustainability and sustainability shifts.
- *Be a knowledge broker*: Scientists have to be knowledge brokers and cross the boundaries between fields and between science and the policy realm (Litfin 1994; Michaels 2009). Michaels (2009) proposes different strategies so as to realize knowledge brokerage for environmental policy that can also imply for sustainability research. The only adaptation that Michaels' (2009) strategies of engaging, collaborating, informing, consulting, matchmaking and capacity building have to undergo is to focus not on finding the problem owner but on finding a sustainability vision.
- *Put sustainability in action*: Sustainability should not be a content-free word in the political jargon. The role of scientists – especially those engaging in policy oriented research or policy advice – in this context should be to inform and aid in formulating policy actions for sustainability to be achieved by different sectoral policies. Sectoral integration of sustainability should be not only a task of administrative bodies from every sector but also a performance indicator of institutional coordination within the government.
- *Create paradigms or lifestyle icons of sustainability*: We suggest scientists to include other actors such as corporate actors, public policy actors and citizens to co-create a paradigm or a lifestyle icon for society to imitate. The trivial mottos of sustainability scientists of "endangering the future of our children" or "borrowing from our children" do not appeal to the citizens. On the contrary it creates strong feelings and anti-movements like the anti-climate change movement. Looking at lifestyle icons that drove societal change (such as the western lifestyle dream or the modern life icons), our suggestion is for scientists to

approach corporate actors and citizens and co-construct either a lifestyle icon or a paradigm that appeals to people that incorporates sustainability principles with values and icons. We believe that if scientists alone create a sustainability paradigm, there is a risk of constructing/formulating utopic images or simply, a utopia. Hence, creating paradigms with a team of actors this risk may be eliminated.

6 Reflection and Discussion

This paper sketched the outlines of the transitions approach on sustainability and the consequences that are drawn from this approach in terms of governance guidelines for sustainability transitions. The transitions approach and transition management focus on understanding and promoting societal processes, and thus, on integrating theoretical with empirical knowledge. We have argued that a process-based approach to sustainability and the integration with the transition perspective has implications for the role of research and knowledge: transitions are uncertain processes that cannot be predicted or fully analyzed. Hence, we argue that certain key patterns and dynamics can be understood and used to reflect upon the possibilities for accelerating and orienting these transitions.

The research needed in understanding and in dealing with transitions is of an inter- and trans-disciplinary nature. Consequently, the research questions formulated cannot be answered in a traditional way: the empirical object (transitions) is continuously on the move. Transition research poses a challenge to the scientific community at large: the complex sustainability problems require the involvement of scientists who step over the boundaries of their scientific disciplines so as to develop new insights, to transfer new knowledge and in general to become part of the collective societal search process we call sustainable development.

Acknowledgements The present paper is a product of research that has been conducted with the support of the KSI II.2 Research Subprogram "The dynamics of transitions" (www.ksinetwork.nl). An earlier version of the present paper had been presented in the EU Conference, Sustainable Development: A challenge for European Research, 28–29 May 2009, Brussels, Belgium.

References

Arthur, W. B., Durlauf, S. N., & Lane, D. A. (1997). *The economy as an evolving complex system*. Reading: Addison-Weasly.

Bijker, W., Hughes, T., & Pinch, T. (Eds.). (1987). *The social construction of technological systems*. Cambridge: MIT Press.

Clark, W. C. (2003). Sustainability, energy use and public participation. In B. Kasemir, J. Jager, C. Jaeger, & M. Gardner (Eds.), *Public participation in sustainability science*. Cambridge: Cambridge University Press.

Clark, W. C., Crutzen, P. J., & Schellnhuber, H. J. (2005). *Science for global sustainability: Toward a new paradigm*. Cambridge: Harvard University Press.

Flyvbjerg, B. (2001). *Making social science matter. Why social inquiry fails and how it can succeed again*. Cambridge: Cambridge University Press.

Frantzeskaki, N., & de Haan, H. (2009). Transitions: Two steps from theory to policy. *Futures, 41*, 593–606.

Funtowicz, S. O., & Ravetz, J. R. (1994). The worth of a songbird – ecological economics as a post-normal science. *Ecological Economics, 10*, 197–207.

Geels, F. (2004). Sectoral systems of innovation to socio-technical systems; Insights about dynamics and change from sociology and institutional theory. *Research Policy, 33*, 897–920.

Gunderson, L. H., & Holling, C. S. (2002). *Understanding transformations in human and natural systems*. Washington, DC: Island Press.

Hendriks, C. M., & Grin, J. (2007). Contextualising reflexive governance: The politics of Dutch transitions to sustainability. *Journal of Environmental Policy & Planning, 9*(3–4), 333–350.

Hisschemöller, M., & Hoppe, R. (1996). Coping with Intractable Controversies: The Case of Problem Structuring in Policy Design and Analysis. *Knowledge and Policy: The International Journal of Knowledge Transfer. 8*, 40–60.

Hisschemoller, M., Hoppe, R., Dunn, W., & Ravetz, J. (Eds.). (2001). *Knowledge, power, and participation in environmental policy analysis*. New Brunswick: Transaction Publishing.

Holland, J. H. (1995). *Hidden order: How adaptation builds complexity*. Cambridge: Helix books/ Perseus books.

Jansen, L. (2003). The challenge of sustainable development. *Journal of Cleaner Production, 11*, 231–245.

Kasemir, B., Jager, J., Jager, C., & Gardner, M. (Eds.). (2003). *Public participation in sustainability science*. Cambridge: Cambridge University Press.

Kates, R. W., Clark, W. C., Corell, R., Hall, J. M., Jaeger, C. C., Lowe, I., et al. (2001). Environment and development – sustainability science. *Science, 292*(5517), 641–642.

Kauffman, S. (1995). *At home in the universe: The search for laws of complexity*. Oxford: Oxford University Press.

Kooiman, J. (Ed.). (1993). *Modern governance*. Newbury Park: Sage.

Kooiman, J. (2003). *Governing as governance*. Newbury Park: Sage.

Litfin, K. T. (1994). *Ozone discourses: Science and politics in global environmental cooperation*. New York: Columbia University Press.

Loorbach, D. (2007). *Transition management: New mode of governance for sustainable development*. Utrecht: International Books.

Loorbach, D., & Rotmans, J. (2010). The practice of transition management: Examples and lessons from four distinct cases. *Futures, 42*(3), 237–246.

Martens, P., & Rotmans, J. (2002). *Transition in a globalizing world*. Lisse: Swets & Zeitlinger B.V.

Meadowcroft, J. (1997). Planning for sustainable development: Insights from the literatures of political science. *European Journal of Political Research, 31*, 427–454.

Meadowcroft, J. (2000). Sustainable development: A New(ish) idea for a new century? *Political Studies, 48*, 370–387.

Michaels, S. (2009). Matching knowledge brokering strategies to environmental policy problems. *Environmental Science and Policy*, Volume 12, Issue 7, November 2009, pp. 994–1011.

Mulder, K. (Ed.). (2006). *Sustainable development for engineers: A handbook and resource guide*. Sheffield: Greenleaf Publishing.

Pezzoli, K. (1997). Sustainable development: A transdisciplinary overview of the literature. *Journal of Environmental Planning and Management, 40*(5), 549–574.

Ravetz, J. R. (1999). What is post-normal science. *Futures, 31*, 647–653.

Rotmans, J. (1994). Transitions on the move; Global Dynamics and Sustainable Development, Bilthoven, The Netherlands: Rijksinstituut voor Volksgezondheid en Milieu (RIVM)
Rotmans, J. (1998). Methods for IA: The challenges and opportunities ahead. *Environmental Modeling and Assessment, 3*, 155–179.
Rotmans, J., Kemp, R., & Van Asselt, M. (2001). More evolution than revolution: Transition management in public policy. *Foresight, 3*(1), 17.
Schumpeter, J. A. (1934). *The theory of economic development*. Cambridge: Harvard University Press.
Scott, W., & Gough, S. (Eds.). (2004). *Key issues in sustainable development and learning: A critical review*. London: Routledge.
Shove, E., & Walker, G. (2007). CAUTION! Transitions ahead: Politics, practice and sustainable transition management. *Environment and Planning A, 39*(4), 763–770.
Teisman, G. R., & Klijn, E.-H. (2008). Complexity theory and public management. *Public Management Review, 10*(3), 287–297.
UN. (1997). *Global change and sustainable development: Critical trends*. New York: United Nations, Department for Policy Coordination and Sustainable Development.
UN. (2005). *The millennium development goals report*. New York: United Nations.
Van Asselt, M., & Rijkens-Klomp, N. (2002). A look in the mirror: Reflection on participation in integrated assessment from a methodological perspective. *Global Environmental Change: Human and Policy Dimensions, 12*, 167–184.
Van de Kerkhof, M. (2006). Making a difference: On the constraints of concensus building and the relevance of deliberation in stakeholder dialogues. *Policy Sciences, 39*(3), 279–299.
WCED. (1987). *Our common future*. Oxford: University Press.
Wildavsky, A. (1979). *Speaking truth to power. The art and craft of policy analysis*. Boston: Little, Brown & Co.

Integrated Climate Governance (ICG) and Sustainable Development

J. David Tàbara

1 Introduction

The present paper introduces for the first time the concept of *Integrated Climate Governance* (ICG) and critically discusses its implications for EU research and policy on 'sustainable development'. ICG is understood as a transition-oriented appraisal approach focused on the creation of assessment tools, policy instruments, and agent-based capacities aimed at dealing in an integrated way with multiple scales and domains related both with mitigation and adaptation. The goal of ICG is to support agent transformation for sustainable development. ICG constitutes both a descriptive and normative synthesis of a large corpus of literature and research within the fields of Integrated Assessment (IA), Integrated Sustainability Assessment (ISA; Rotmans et al. 2008), Social and Sustainability Learning (Pahl-Wostl et al. 2008), and research on the institutional dimensions of global environmental change (Young 2008).

The reflections provided in the following pages derive mainly from the insights gained from my involvement in several EU funded research projects, most recently MATISSE (Methods and Tools for Integrated Sustainability Assessment; www.matisse-project.net), and ADAM (Adaptation and Mitigation Strategies – Supporting EU Climate Policy; www.adamproject.eu), as well as in the new IHDP project Integrated Risk Governance (Jaeger et al. see http://www.irg-project.org/) and Global Systems Dynamics and Policies (www.gsdp.eu). However, and for reasons of space, specific references to the empirical material of these projects have been omitted here and the following contribution concentrates only on the more analytical, operational and normative aspects of the concept. It is argued that ICG constitutes a powerful conceptual synthesis to reframe present EU research and policy making processes and outcomes on climate change in ways which increasingly become more relevant to

J.D. Tàbara (✉)
Institute of Environmental Science and Technology, Autonomous University of Barcelona, Campus UAB. E-08193 Cerdanyola del Vallès, Barcelona, Catalonia, Spain
e-mail: Joandavid.tabara@uab.cat

C.C. Jaeger et al. (eds.), *European Research on Sustainable Development*,
DOI 10.1007/978-3-642-19202-9_8,
Published by Springer-Verlag Berlin Heidelberg 2011

meet pressing societal and policy needs and to support goals such as the ones stated in the EU Sustainable Development Strategy.

A main goal of this chapter is to stimulate a discussion on how and by what processes current EU research and policy on climate change could be reframed in a way which can best support the design and implementation of sustainable development at different levels of governance. However, I do not attempt to define what the ultimate meanings of governance or of sustainable development are (Jordan 2008), discuss the seriousness of the current climate change risks or evaluate the adequacy of the existing climate assessment processes (Jaeger et al. 2008). The focus here is to propose a heuristic device to help to integrate the science and policy of climate change into a common framework with the explicit goal of stimulating agents' transformations and institutional innovation. Progress in sustainability, rather than in 'sustainable development' as the concept in itself can be thought as a trap (see M. Midgley in this volume) is more likely to be made by addressing directly a specific number of urgent persistent problems -and their interrelationships- than by trying to agree on its ultimate semantic definition before it can be implemented. At the very end, what is needed is systems to move towards highly diverse sustainable patterns of social-ecological interactions, and not a single definition of sustainable development to be applied in all contexts. From this perspective, the starting point is that climate change not only poses one of the greatest challenges to humankind has to urgently respond but also constitutes one of the greatest opportunities for making real progress in sustainability.

2 Climate Change as a Large-Scale Persistent Problem

Persistent Problems can be defined as the class of problems for which ultimate causes have to be found in the previous application of inadequate solutions to deal with them. Persistent problems often emerge as a result of applying simple, singled-perspective and linear solutions to systemic processes of socioecological change. The intensity and scale of persistent problems increases when no institutional and social learning occurs. The choice of one-dimensional measures for problems and policies which are inextricably interlinked together – such as those related to water scarcity and pollution, energy, land-use management, or biodiversity/ecosystems functions conservation – often result in the accumulation of negative side effects and the worsening of the initial conditions of the systems of reference in which such problems originally emerged (Fig. 1).[1]

Unless properly framed and managed, climate change has the potential to enter into this class of problems, whereby recursive, cumulative and eventually intractable feedbacks express themselves in multiple ways and at different levels of

[1]In contrast to Young (2008:124) I understand that cumulative environmental problems are those which their feedback effects become forces of environmental change by themselves. In this regard, climate change is both a cumulative problem and a systemic one.

Fig. 1 The growth in scale and intensity of global environmental related risks is often the result of the accumulation of feedbacks derived from intricate environment-societal interactions and decisions

action. A usual way to deal with persistent problems and risks in the short term at the local and regional level is to try to shift their most visible negative effects and costs to other contexts, scales, or policy domains without actually eliminating their original causes. Adequate policies to deal with climate change need a holistic and integrated approach, which takes into account the systemic, random and multiple feedbacks – positive and negative – derived from both social and ecological systems and their interactions.

3 Power, Knowledge Management and Social-Ecological Transitions

Power is inextricably embedded in all the processes which concern the scientific definition of the potential impacts of climate change, the selection of policy measures, and the communication of the messages to be disseminated to the public. Power conflicts lay at the heart of the boundaries management which mark who can participate in the definition of the problem and of the possible alternatives of action. Decisions on risk assessment, policy instruments, as well as on communication and learning are always embedded in particular power structures. Transitions in the management of large scale risks such as global warming inevitably require modifications in the distribution of power, changes which in turn are materialised in new forms of institutions.

In addition, very often the costs and benefits of persistent problems and of their potential solutions are distributed unevenly across different sectors of the population. Inequality in the access and use of natural resources and in the processes of knowledge production limits the scope of agent and system transformations. The resistance by the incumbent regime to resolve such inequalities is often the main obstacle which impedes the type of transformations which would lead to a long-term structural improved situation. Dealing with large-scale complex risks requires empowering specific agents and to support niche developments in particular contexts of action – mainly at the local and regional level – in a way which enables them to break the existing lock-in situations and participate in the configuration of new regime institutions. In particular, the making of new local and regional institutions capable of dealing in a social-ecological robust way with large scale environmental risks demands the redistribution of power at different levels.

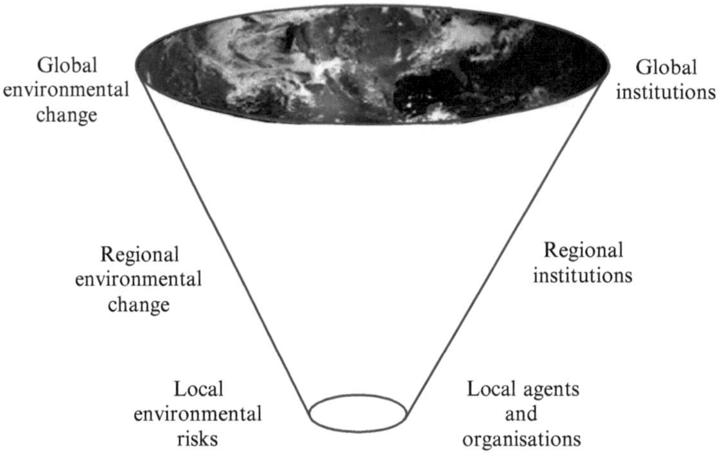

Fig. 2 Multi-level social-ecological dynamics and governance. The governance of large scale environmental risks such as climate change demands a clear definition of responsibilities as well as coordination between different levels of policy action to create incentives and conducive conditions for agent transformation

Transitions in climate change governance require *systems' learning,* although this cannot be achieved without empowering agents at lower levels. From the field of Integrated Sustainability Assessment (ISA Rotmans et al. 2008) it has been clear that unless new processes, tools and methods are developed which are specifically addressed to tackle issues of power and agent transformation, there is little change of progress toward a transition to sustainability. Policies aimed at supporting transitions, such as the one that would eventually lead to a progressively decarbonised global society, need conducive institutions which define clear responsibilities at different levels, while at the same time can provide adequate mechanisms for coordination between each of them (Fig. 2).

4 A Systems Perspective on Agents' Transformations and Complexity

Nevertheless, the possibilities and capacities of individual agents and organisations to cope with multiple risks and modify their behaviours accordingly depend on the particular interactions they maintain with the structures of the overall social-ecological system in which they operate. Such structures are composed by social governing rules but also by biophysical constraints which characterise their ecological base, and this provides the potential as well as the limitations for agents' transformation, learning and reconfiguration. And in this way, different patterns of social-ecological system' configuration may be more or less conducive to adapt or mitigate problems of unsustainability.

To underline these points we can use the integrated meta-modelling framework SEIC (Tàbara and Palh-Wostl 2008) – in which all socioecological systems and agents dynamics can be understood as resulting from the interactions between their structures (S), the use of energy and resources by these agents (E), the available information and knowledge (I) and the accumulated socio-ecological change (C) that their activities produce. According to this perspective, within a particular system level, a greater availability of resources, new access to knowledge and information, or new governing rules may create new degrees of freedom or new spaces for agent interaction which may facilitate agent transformation, social learning, and eventually, increase the potential for transition. In contrast, a situation in which there is an inflation of rules, an excess of physical interactions, or where information or the availability of resources becomes relatively scarce may lead to congestion, and in these conditions agents' transformations – and the overall transition – may become more difficult if not impossible to achieve if this process is not accompanied with an increased in the overall complexity of the system (Fig. 3).

World natural resources, information systems and institutions are increasingly interconnected with each other, thus increasing their interdependency and inter-configuration. However, when such structural connectivity does not follow a modular shape (e.g. where individual parts of the system cannot be 'disconnected' or allow for self-organisation or self-configuration) this may create new conditions for a higher vulnerability to small changes and shocks occurring in any other parts of the system. As a working hypothesis, we can say that the capacity of a society to

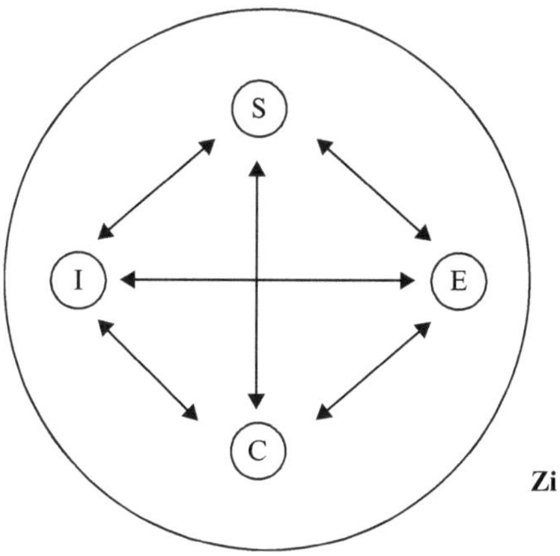

Fig. 3 The social-ecological system as characterised by the SEIC Model (Tàbara and Palh-Wostl 2008). Whereby S: Structure, E: Energy and resources; I: Information and knowledge; and C: accumulated socio-ecological change; zi relates to the size of the system of reference and depends on the selected unit for observation and analysis. Social-ecological systems are composed by agents that respond to these four types of subsystems constraints and dynamics. Transitions in climate governance and sustainable development demand profound changes in the configuration and dynamics of each of these four subsystems which affect agents' behaviours and interactions between them

cope with large-scale persistent risks, such as climate change, and of its agents to transform themselves accordingly, depends on the level of structural complexity achieved by S in ways that minimises the use of E, reduces negative C, and does so by the best use of I. This dynamic can be referred to as an increase in 'sustainability oriented system complexity'.

In this guise, the 'normal' evolution of all SESs leads to congestion and entropy once the degrees of freedom available to the agents that conform these systems are exhausted. The reasons for that can be manifold, and can derive from an excess in the use of resources and energy (E) based on existing technologies, the accumulation of its negative side effects (C), or even the lack of adequate rules (S; or the excess of them) that impede agents' transformation, increased number of biophysical interactions, and/or the possibility to develop or use new sources of knowledge and stocks of information (I; e.g. embedded in diversity). All in all, a central tenet is *that the sustainability of a social-ecological system depends essentially on the quality of the information and knowledge systems used by the agents who form the overall socio-ecological system, and in particular to the extent such information and knowledge support adaptive transformations of existing structures*. Therefore, sustainability transformations rely on the capacity of agents to develop and use information and knowledge systems to accurately in modes that describe existing dynamics and activate feasible options and alternative pathways of development and growth. This can only be done whenever such development also produces a sufficient number of degrees of freedom for the agents to be able to reorient their behaviour without limiting their possibilities of quality interactions (not necessarily physical) and learning.

Thus, in principle, the more complex a society is structured, the more capable may it be to deal with complex problems, but only if such complexity contributes to meeting the above criteria of progressively reducing the overall negative systemic change (C) by making the best use of its information and knowledge systems (I). Indeed, new persistent problems may derive from the development of an increased complexity, as a more complex society also tends to demand greater levels of information, energy, and resources and provoke greater environmental change to maintain its structure. So the key is *how* such complexity is used and for what purposes, in the sense that an increased structural complexity is a necessary condition, albeit not a sufficient one, to deal with the increased complexity of the current problems of unsustainability. In this sense, the level of structural global complexity is also closely related to the preservation of global diversities, that is, natural, institutional and cultural, as these constitute the repositories of alternative sources of knowledge as of rules which are needed for sustainability learning. If these assumptions hold true, the governance of highly complex problems such as climate change demands not only the management of complexity, but also explicit policies and measures to increase and reorient it in the right direction to confront the social-ecological systemic effects derived from it. Indeed, the success or failure of different pathways aimed at managing sustainability and risk transitions may therefore be dependent on the *level and type* of complexity achieved by different social-ecological systems, a set of properties which can only be developed from social learning (Fig. 4).

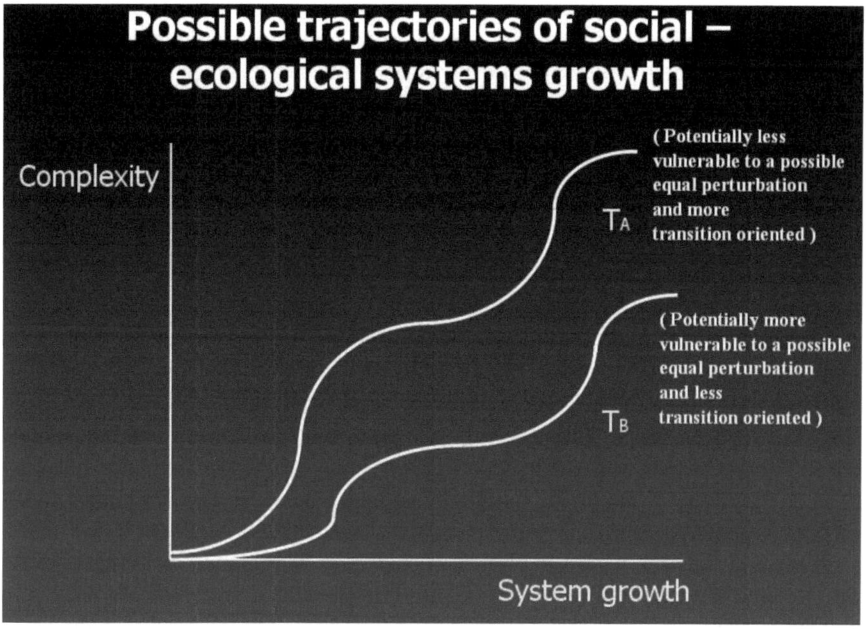

Fig. 4 Complexity growth and possible social-ecological adaptive trajectories within a particular system level, the long term management of large scale environmental risks such as climate change needs explicit policies aimed at increasing and reorienting complexity, a process which can only be achieved as a result of social learning, which is mostly an institutional challenge[2]

The existing power and governance structures also condition the type of policy choices which are generally favoured or discouraged by the existing regime, which in turn also affect the options and pathways for resource transitions. Overall, policies based on ecoefficiency tend to be encouraged by the existing regime because they allow for an increase in the options for resource use and social interactions – thus postponing system congestion – without questioning much the existing regime. At the same time, such policies and measures may have few negative distributional effects in the short term, hence lessening the potential for conflict. However, policies based on managing sufficiency, that is, setting limits to resource use, often entail new compromises between existing power forces. Setting maximum thresholds to resource use or pollution levels – such as reducing current GHG emissions to about 80% or more below 1990 levels by 2050 – would demand profound transformations in agents' practices and system reconfiguration. This may threaten the existing power regimes and would require the empowerment of new niche coalitions in order to achieve and reduce the impact of its potential

[2]This does not mean, however, that all societies need to follow the same structuration pattern, but on the contrary, that many different patterns and configurations and organisation are required – and not only one – in a more complex society which aims to cope with the challenge of growing unsustainability.

distributional costs. Needless to say, an integrated approach to climate adaptation and mitigation in tune with the predicament of sustainable development requires of both types of policies, dealing as much with ecoefficiency as much with sufficiency.

5 Towards Integrated Climate Governance. Concept and Applications

A common reaction by regime agents when confronted with the increasing effects and interdependency of multiple systemic constraints in the fields of energy, land use, food production, water, and climate change is to respond with non-systemic and non-integrated responses, thus trying to look for single indicators of success or failure (Fig. 5). However, simple solutions to the complex problems of unsustainability are no longer a possible answer. Such widely shared cultural attitude impedes not only dealing effectively with climate mitigation and adaptation but also reaching a much broader system transition to a more secure and sustainable development based future.

However, and with regard to climate change, it is increasingly clear that policies dealing with the type of Green house Gases (GHG) reductions which are needed to prevent catastrophic losses in the relative short term cannot be assessed or managed in isolation. To prevent problem shifting and rebound effects, what is needed is the development of a set of integrated multi-domain, multi-scale approaches, each one able to take into account these multiple constraints and ecosystem limits and to support transition and transformation processes at different system governance levels.

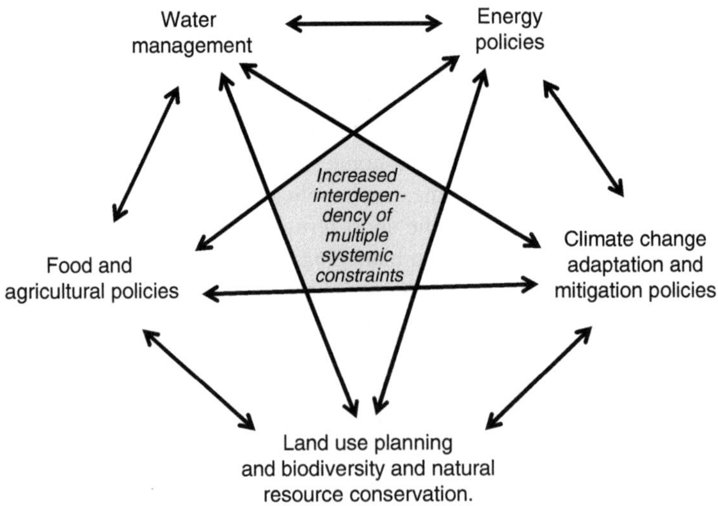

Fig. 5 The growing interaction between multiple constraints and policies, demands a multi-domain, multi-scale integrated science-policy-communication approach both in science as in policy with interaction with the public. New policy domains and new ecological threats are being acknowledged and added to this complex 'treadmill of social-ecological constraints'

A possible and feasible way to start doing so is by promoting basic innovations in the organisation of time, space and resource use – e.g. using different tools and mechanisms for learning and behavioural change to reorient agents' interactions and structural social-ecological configurations. These procedures should be able to support the setting of voluntary quantitative limits to natural resource use so to avoid rebound effects which emerge from policies which solely focus on improving ecoefficiency. In our democratic liberal societies, such 'voluntary management of austerity' – which in itself can be a source of sustainability innovation – needs necessarily to be carried out in a collective and participatory way. Therefore, an integrated governance approach which acknowledges and anticipates ecological limits should not only be interested in learning about 'what to do' but most notably also 'what not to do' – some of the most difficult things to learn in our opulent liberal societies. Similarly, this approach would need to reframe the current socio-ecological challenges not as problems about what to do with 'the environment out there' but mostly as a challenge about what to do with our interactions with the total socioecological system – hence overcoming some of the persistent cultural dualisms that dominate our perception of the total environment and which are so much ingrained in the obsolete distinction between the 'social', 'ecological' and 'economic' realms. In this regard, not only changes in power are needed and have to be promoted, but ultimately in culture. It is through the lens of culture which we define the basic dimensions of time, space, and happiness and which determine our interactions with the natural world in more or less sustainable ways.

At present there is no such set of integrated appraisal approaches able to assess climate risks from this integrative holistic perspective and to support societal transitions and agents' transformations in different governance contexts (for a conceptual attempt, see Weaver et al. 2006). In policy, the recursive negative effects between different policy goals and measures, which often follow contradictory purposes framed at different time and spatial scales, are generally assessed or dealt with separately. This situation leads to a continuous practice of problem shifting rather than problem reframing, which transposes, rather than solves, the different externalities either to different scales or to different domains.

The concept of Integrated Climate Governance (ICG) aims to fill this gap. ICG constitutes a theoretical and epistemological synthesis of a large corpus of scientific literature on Integrated Assessment, Integrated Sustainability Assessment, Social Learning, Risk Analysis, and research on institutions of Global Environmental change as well as climate change appraisal. It is intended to set up a research programme and to help analyse existing practices and to provide guidance for scientists and policy makers, as well as practitioners working in the interface between the two (media, knowledge brokers, civic society organisations) regarding climate action. Therefore, the present framework of analysis does not consider or focuses only on the improvement of the tools and methods for the scientific assessment of climate risks, nor only on the possible policy architectures and instruments, or solely on the processes of public communication and outreaching. *The emphasis of ICG lies precisely in the learning and transformation processes which result from the interaction and integration between the three spheres,* and not

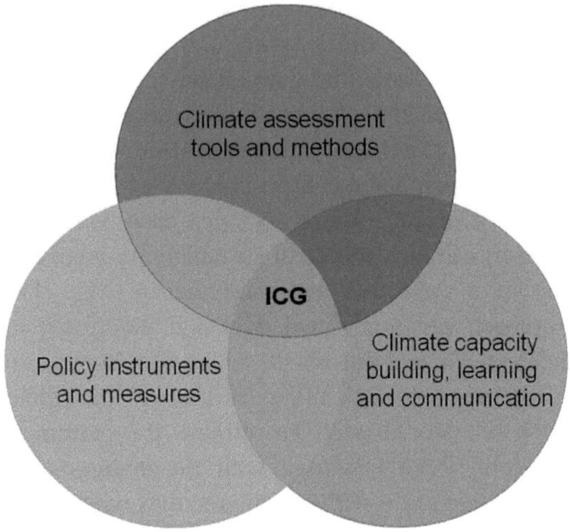

Fig. 6 Integrated climate governance

in any one of them alone (Fig. 6). Future research on ICG may concentrate on examining what type of mechanisms foster new forms of interaction and reconfiguration of agents and institutions working within and between these three spheres. The normative goal being to support the design and implementation of efficient, equitable and social-ecologically robust climate strategies in multiple contexts of application in synergy with other strategies such as the implementation of the EU Sustainable Development Strategy.

In short, Integrated Climate Governance (ICG) can be defined as:

> The structured generation and use of tools and methods that combine a plurality of legitimate but divergent interests and sources of knowledge and judgement for: (i) the comprehensive assessment of climate risks and opportunities, (ii) the design and implementation of policy instruments, and (iii) the creation of communication, engagement and transformative learning capacities, all aimed at producing long-term efficient, equitable and socially and ecologically robust climate strategies. ICG deals both with adaptation and mitigation, and does so from a multi-scale, multi-level, multi-domain and transition-oriented perspective.

Therefore, ICG is as much a process of transformative assessment (appraisal) as it is a process of governance and public communication and learning. The goals and strategies of ICG will never be predetermined but will result as an outcome of a social learning process derived from the interaction between: (1) The actual assessment procedures on risks and opportunities (2) The development of policy instruments and their implementation and (3) Communication and capacity building. ICG needs to integrate multiple sources of local and universal knowledge and judgement, not only to deal with uncertainty but also because the necessary engagement of agents and the understanding of the systemic conditions which determine their potential transformations can not be achieved only by top-down strategies. The emphasis on communication and public awareness is crucial, given that at present most dominant information systems (including the media and the market price systems) are still largely oblivious to the reality of climate and unsustainability threats, while these conform the basis for agents' and system transformations. ICG could be particularly useful to orient current research on climate change in a way which becomes more relevant to support policy decisions.

In particular, a context-based local and regional approach to Integrated Climate Governance would:

- Emphasise the spatial dimension in the appraisal of climate mitigation and adaptation options and strategies. It should apply a multi-scale approach to the design of climate strategies, hence looking at global and local processes as well as taking into account the long-term needs and present opportunities for institutional and agent transformation. The role played by land use planning is central in local and regional ICG.
- Integrate and combine the assessment of vulnerabilities with the development of opportunities for business and regional sustainable development. Regional ICG may not only be concerned with creating capacities to reduce the potential impacts upon the most potentially vulnerable populations, but above all with developing an adequate set of incentives, social networks and policy instruments mix capable of stimulating transformations and turning potential risks into opportunities. Thus, a central aspect of ICG is its special focus on the development of a portfolio of incentives and feasible options which take into account trade-offs as well as synergies between multiple domains, while considering the particularities of different agents and sectors demands and needs.
- Be based on the continuous identification and empowerment of local niche developments and successful experiments both in adaptation and mitigation, with the goal of being up-scaled and mainstreamed into larger regional or international programmes.
- Develop efficient mechanisms and capacities to help vertical and horizontal institutional coordination. An important task in this regard is to increase the complexity of the institutional climate arrangements and to support the development of new cross-cutting learning networks which facilitate the integration of assessment tools, policy instruments and knowledge transfer between the climate domain and other policy domains.
- Fully apply the precautionary and subsidiary principles to guide climate decisions at the local and regional levels. In this sense, ICG would aim at developing mechanisms which take into account the uncertainties about the local and regional climate impacts and of the effectiveness of possible measures to cope with them, in line with approaches close to adaptive management. ICG research and policy at these lower levels would encourage and support adaptation and mitigation initiatives which go beyond the mere compliance with national and international GHG reduction targets thus opening up new spaces for interaction, experimentation and learning among local and regional agents.
- Aim at integrating equity, efficiency and diversity in the design and selection of climate policy options and strategies. ICG would not only look at the potential

results of decisions but also at the processes and initial conditions in which the decision making processes are carried out. It would also consider not only the contribution of distinct policy options to specific targets of climate mitigation and adaptation but especially how the attendant costs and benefits are distributed among the affected populations. ICG designs need to be flexible enough as to accommodate new distributional and efficiency criteria on the basis of new policy values and scientific knowledge.

The above general framework proved useful to identify, analyse and understand what is still missing and what has already been achieved with regard to the development of transformative and robust appraisal capacities in the ADAM selected case studies[3]. New capacities in the three main spheres of ICG – that is, in risk/opportunity assessment, policy implementation, and social learning and communication – were observed and developed in the three regions. Nevertheless, and given that the social, ecological and institutional contexts in these contexts are so diverse, these capacities and how they were integrated into the making of a regional strategy differ greatly. A prominent result of the comparative analysis of the ADAM regional case studies is that the level of integration between these three domains was on the whole very low, even though there are venues and opportunities to make such integration possible. Furthermore, the type of science used for assessing risks and opportunities is not sufficiently well-equipped. At present prevailing science practice is largely unable to provide integrated narratives on such interconnected and systemic problems of human-induced climate change and unsustainability in ways that are relevant for policy.

6 Making ICG Operational at Local and Regional Levels

The concept of Integrated Climate Governance has been introduced as a heuristic devise which can be used to identify the gaps and potentialities of existing climate appraisal practices from a multi-scale, multi-domain and comprehensive transition-oriented perspective. It can also be used to support large EU sustainability policy processes such as the implementation of the EU Sustainable Development Strategy. This is so, because ICG can also be used not only to reframe climate policy making but also policies oriented at provided new forms of science practices in line with the transformative approaches defended in this book. Such broad perspective, however, requires operationalisation. One option to do so is to provide an initial set of questions which can trigger the necessary discussion on how to develop a more coherent and consistent development of climate appraisals tools, policy processes and communication strategies whenever required. To link ICG to local and regional sustainable development goals, the following questions may apply:

[3]Which are the Tisza floodplain in Hungary, the Guadiana river basin in Iberia and the Inner Mongolia region in China; (www.adamproject.eu; Tàbara 2010; Tàbara et al. 2010).

1. Risk assessment:
   - To what extent do existing assessment tools and methods deal with feasible options for social-ecological system transition, agent transformation and institutional collaboration?
   - To what extent do climate assessment procedures look at both positive and negative effects derived from changing institutional arrangements and distribution of responsibilities?
   - What type of knowledge is needed to improve both adaptation and mitigation capacities of agents at the local and regional level, beyond the representation of potential impacts? (e.g. on incentives, options and institutional reforms)
   - To what extent do existing tools and methods used in the assessment of climate risks take into account the local and regional institutional constraints and potentialities which impede or facilitate both adaptation and mitigation?
   - To what degree do the criteria for the selection of tools and procedures for assessing local and regional climate risks and opportunities take into account multiple scales, domains and the potential role of multiple governance levels?
   - How do different appraisal tools and methods used by regional agencies employed to devise their climate strategies differ, produce synergies or contradictions from those used at the national or European level?
   - To what extent and how do regional appraisal agencies downscale and integrate global scientific knowledge about climate impacts to design and implement their own adaptation and mitigation strategies? And in particular, what is the level of plurality in the use of tools, methods and procedures used by local and regional agents to downscale global assessments and insights?
   - To what extent and how are local and regional perceptions, experiences, needs and opportunities regarding adaptation and mitigation up-scaled to the international appraisals and governance processes?
2. Policy practice and implementation:
   - To what extent local and regional measures on mitigation and adaptation are mainstreamed within and broader strategies regarding sustainable development? And to what degree is such climate mainstreaming understood as an indicator of progress in sustainable development?
   - What changes can and are being carried out within the local and regional institutional arrangements in order to integrate climate change and sustainable development strategies in their present organizational structures?
   - To what extent the design and implementation of new climate policy instruments and strategies not only include the best knowledge about the climate dynamics and impacts but also address issues of inequality, different power distributions and processes of policy reframing and learning?
   - What is the potential contribution of local and regional agencies in devising climate strategies that go beyond the compulsory compliance of international commitments?

- What is needed for local and regional governments to develop and implement long-term climate policies which integrate and create synergies between adaptation and mitigation strategies?
- What is the influence of the existing national governance structure in constraining or enhancing the development of local and regional capacity of agents to intervene in climate policy? And in particular, what is the influence of decentralised and polycentric governance structures in this regard?
- What type of synergies and trade-offs can be observed with the implementation of climate policies and measures and the implementation of policies and measures in other domains?
- Do current climate and sustainable development strategies consider structural systemic issues to reduce vulnerabilities and enhance ecological resilience such as to promote modular connectivity, setting voluntary limits, or avoiding congestion of social-ecological systems?

3. Communication, social learning and transformative capacity building:

- What new networks and new spaces for interaction and collaboration between science, policy, and the public can be developed to stimulate learning, transformation and long-lasting integrated capacities on climate and sustainable development?
- To what extent does communication on climate change adaptation and mitigation need to be carried out in different ways and languages among different audiences to exert the expected influence (Mitchel et al. 2006)?
- How can system feedbacks and processes – rather that solely 'impacts', 'problems' and 'solutions' – best be communicated to wider audiences and by what means to encourage societal change and engagement?.
- How can 'communication' and 'participation' be turned into durable agent engagement and transformation?
- How and to what extent such public engagement can become part of the assessment and knowledge production processes to increase their salience, credibility and legitimacy (Jäger and Farell 2006)?
- What are the particular knowledge needs and required assessment capacities of agents to make ICG relevant at the local and regional level?
- How new scientific insights on the current state of the climate and of the institutional and social system are incorporated in local and regional climate appraisals and strategies?
- To what extent do current communication processes address issues of cultural reframing, in a way that aim to overcome the prevalent dualisms between biophysical and human systems? (including the obsolete distinction between 'social', 'ecological' and 'economic' realms)

The concept of ICG derives from the realisation that an important gap exists between the processes of knowledge building and risk/opportunities assessment, the design and implementation of policy instruments and the communication and

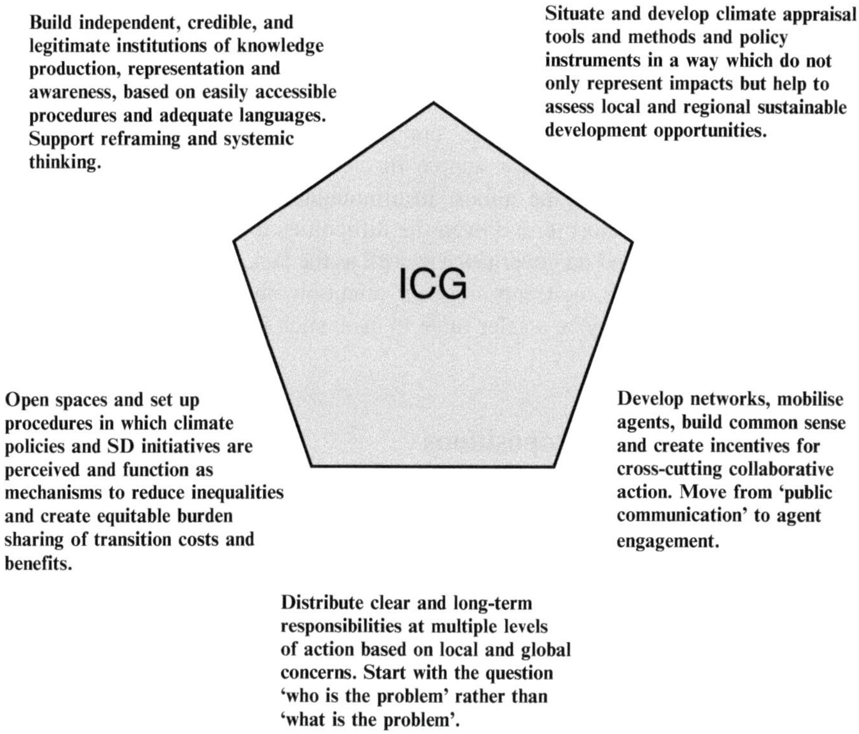

Fig. 7 Five basic interlinked tasks to support Integrated Climate Governance

societal responses regarding climate change. New institutional integrated mechanisms need to be created capable of linking these processes which occur at the global level with those that are undertaken by local and regional agents and vice versa. In order to do so, and from the insights gained within the MATISSE and ADAM projects, a series of key basic tasks have been identified which could help to move current regional appraisal practices towards a more ICG approach, as represented in Fig. 7.

Therefore, an approach based on governance, rather than only on 'government', or that moves beyond solely producing 'more' knowledge, expert tools and methods, would emphasise the need for articulating long-term institutional mechanisms to enhance the quality of *the interaction, the type of knowledge exchanges and the learning processes* between climate scientists, policy makers and the general public – eventually in charge to take the necessary adaptation and mitigation actions. Questions such as who participates, through what institutional mechanisms, or what tools and methods are being used to create climate knowledge and for what purposes, take a different dimension when framed through the relational socioecological perspective of Integrated Climate Governance. This is an institutional challenge that demands new and more complex forms of system configuration as well as of spaces for transformation in a way that the negative systemic effects of the actual growth of socio-ecological systems are also taken into account.

In the future, the scope and normative goals of ICG could be extended to include broader issues of sustainable development in general, thus becoming a plausible 'Integrated Sustainability Governance' approach. However, while such an overarching approach could be a new source for science and policy innovation, at present, one can only imagine almost insurmountable difficulties to undertake such endeavour in the short term. Given the difficulties to delimit the discussions both on sustainability and on governance as well as the lack of tools and methods to do so, in the meantime, concentrating our attention on climate transformative targets and capacities may be a safer route to meet such a challenge.

7 Conclusions and Propositions

The present paper has attempted to define and begin to make operational a new concept, *Integrated Climate Governance*. In particular, I have argued that ICG can be potentially very useful to reframe and reorient future practices occurring in the intersection of science, policy making, and communication by providing a more coherent analytical and policy-relevant perspective. The ICG concept is based on a particular understanding of the key challenges which regard the adequate governance of environmental change. Mostly, that we need to move from thinking in terms of 'problem-solution' towards a new mode of science and policy practice based on assessing, managing and communicating cumulative feedbacks and interlinked processes of total socioecological change.

This concept is also based on the acknowledgment that actions aimed at developing adaptive capacities at the individual or local level may have negative impacts on mitigation. Whenever responses are not conceived in a holistic way, these are likely to create new problems in new situations, at different scales or domains. Flexible, context-based, and multi-level adaptive processes, as well as new tools and methods are required (Tàbara et al. 2010). Such new global approach to climate appraisal and action should be able to appraise both climate risks and opportunities, deal in an integrated way with mitigation and adaptation, and support the making of robust climate strategies at different governance levels by taking into account multiple synergies and trade-offs between various policy domains, sectors and scales. ICG is set up as a research programme that recognise the need to prevent problem shifting from one policy domain to another. If implemented, ICG should also be conducive to triggering fast transformations in individual practices and institutions, and the building of new partnerships, and the emergence of new framings, new mindsets, and of new spaces for interaction between science, policy making and the public. This is precisely what distinguishes this approach, based on *appraising* from those based on 'assessment': it attempts to go beyond simple representation (e.g. of impacts and trends) and explicitly to support system transition and agents' transformations.

In this sense, the practice of ICG could contribute to changing some of the dominant mindsets and cultural frameworks which are now being used to conceptualise the climate problem, e.g. from a cost/benefit calculation to a problem of governance that requires transition and new forms of global cooperation. Furthermore, ICG offers one of the best chances to reframe international relations, and in this sense to contribute to both scientific and political innovation globally, and in Europe, to ensure meeting its sustainable development goals in the long term. Such reframing entails moving away from the present market-based global competition towards a more sustainable development/climate global cooperation. And in turn, in such global cooperation may lay the best changes for global development and new forms of growth.

From this standpoint, conclusions are presented in the form of four basic normative propositions which future research will need to further elaborate:

1. *Integrated Climate Governance demands above all, institutional innovation, not only more 'tools' and methods.* In order to mainstream climate concerns in sustainable development strategies, new institutions which deal in an integrated manner with risk/opportunities assessment, policy-making, communication and capacity building are needed. Simply 'more tools', 'more instruments' or 'more public communication' alone (following business as usual frameworks and practices) may have little effect on improving climate adaptation and mitigation and contributing to sustainable development. The three spheres of ICG need to be integrated and coordinated into a common facility. The new ICG institutions could help frame more adequately the design of climate assessment tools and methods, the type of policy goals which are being pursued, and to provide accountable procedures for the incorporation of scientific knowledge and public demands in the development of new policy instruments and measures.
2. *Institutional innovation regarding climate change entails enhancing and reorienting the overall complexity of the international and regional systems of climate risk/opportunities assessment, governance and communication while at the same time ensuring global coordination and local engagement.* A more complex but also different institutional landscape is required to meet the type of transitional changes which would deal in a systemic way both with adaptation and mitigation. A broader participation and engagement of regional and local actors in the global challenge of climate change demands the development of new capacities, tools and methods for assessing regional impacts of climate change, of more locally and regionally adapted policy instruments, as well as regionally suited mechanisms for communication and learning. But at the same time, overarching coordination mechanisms are also required within the three spheres of ICG. Institutional innovation is needed regarding the sharing of scientific knowledge as well as in relation to policy architectures and incentives which facilitate public engagement. Robust strategies for climate change demand taking into account the socio-ecological and political specificities of local and regional contexts in a way that explore and enhance the possibilities for a growing social and institutional complexity and coordination.

3. *Meeting significant transformative climate mitigation targets and building adaptive capacities may constitute some of the most decisive ways to support and account for progress on sustainable development and avoid relativism.* Many practices in science, policy and communication are excessively dominated by a strand of social-constructivist, non-integrated and relativistic discourses of sustainability. Such relativistic approaches are often understood as a sign of impartiality, objectivity or scientism while in fact what they are is often precisely the opposite: they unveil the lack of robust knowledge – or the strength of vested interests to use this pretended lack of knowledge and instrumental uncertainty – to provide well-grounded, falsifiable, and transferable insights about what actually works or does not work regarding sustainability. The relevance of science to support decision making on sustainable development largely depends on the capacity of achieving a minimum consensual agreement on how to quantify – and qualify-the contested concept of sustainability[4]. Setting long-term transformative targets for the reduction of GHG emissions at multiple governance levels and monitoring their implications for the short term – e.g. regarding technological and societal innovation-, may avoid some excessive social-constructivist discourses which make the measure, design and implementation of adequate tools and methods to support progress in sustainable development almost unattainable.
4. *The development of new ICG policy instruments needs to incorporate socially and ecologically robust systems of knowledge in the development of long-term transition goals.* The integration of existing scientific knowledge should not only be oriented to support one-shot decisions about the implications of implementing particular regulations but be oriented to help devise processes which already incorporate combinations of policy instruments with coherent transformative system goals. Knowledge about the stocks and flows of ecological systems as well as their dynamics needs to be incorporated into the development of new policy measures. New instruments and processes for climate appraisal should be developed not only to address aspects which regard the administrative boundaries of a single policy domain or scale but should be included in a broader transition framework able to trigger a cascade of innovations and decisions in multiple domains. If the principles of ICG are perceived robust enough as to inform and improve current climate appraisal and policy making process, new instruments need to be oriented to support the transformation of routines and practices in a way which make them more in tune with the current biophysical trends and societal needs. This demands taking a more systemic, adaptive and relational approach to climate appraisal and climate capacity building, able to deal with diversity as well as with the new opportunities for social and institutional innovation. Particular climate instruments ought to be

[4]The recent experience at the EU level – with the Impact Assessment procedures and the EU SDS – shows that to a large extent, the failure to produce a robust and systematic procedure as well as a set of convincing tools and methods to assess sustainability progress relates to a large extent to the difficulty of finding an alternative – and equally powerful-measure to that of GNP.

devised and new independent and credible institutions need to be created for generating new patterns of societal – ecological interaction synergetic with the predicament of sustainable development.

Acknowledgements This paper and reflection could have never been made without the inspiration and intensive years of EU research with Alex Haxeltine, Jill Jager, Carlo Jaeger, Claudia Pahl-Wostl, Jan Rotmans, and Paul Weaver. I am also grateful to the reviewers of an earlier version of the paper and to the colleagues and researchers of the MATISSE and ADAM projects. Responsibility for any possible bias or flaws in the arguments presented is only mine.

References

Jaeger, C. C., Schellnhuber, J. H., & Brovkin, V. (2008). Stern's review and Adam's Fallacy. *Climatic Change, 89*, 2007–2218.

Jäger, J., & Farell, A. E. (2006). Improving the practice of environmental assessment. In A. E. Farrell & J. Jäger (Eds.), *Assessments of global and regional environmental risks. Designing processes for the effective use of science in policy making* (pp. 278–293). Washington, DC: Resources for the Future.

Jordan, A. (2008). The governance of sustainable development: Taking stock and looking forwards. *Environment and Planning C: Government and Policy, 26*, 17–33.

Mitchel, R. B., Clark, W. C., & Cash, D. W. (2006). Information and influence. In R. B. Mitchel, W. C. Clark, D. W. Cash, & N. M. Dickson (Eds.), *Global environmental assessments. Information and influence* (pp. 307–338). Cambridge: MIT Press.

Pahl-Wostl, C., Mostert, E., & Tàbara, J. D. (2008). Special issue on social learning in water resource management. *Ecology and Society, 13*(1), 24. [online] URL: http://www.ecologyandsociety.org/vol13/iss1/art24/.

Rotmans, J., Jager, J., & Weaver, P. (2008). Special issue on integrated sustainability assessment. *International Journal of Sustainable Development and Innovation, 3*(1/2), 1–162.

Tàbara, J. D. (2010). Integrated climate governance in regions? Assessing Catalonia's performance using the 'climate learning ladder'. *Regional Environmental Change.* doi:10.1007/s10113-010-0135-3.

Tàbara, J. D., & Palh-Wostl, C. (2008). Sustainability learning in natural resource use and management. *Ecology and Society, 12*(2), 3. [online] URL: http://www.ecologyandsociety.org/viewissue.php?sf=28.

Tàbara, J. D., Dai, X., Jia, G., McEvoy, D., Neufeldt, H., Serra, A., et al. (2010). The climate learning ladder. A pragmatic procedure to support climate adaptation. *Environmental Policy and Governance, 20*, 1–11.

Weaver, P. M., Haxeltine, A., van de Kerkhof, M., & Tàbara, J. D. (2006). Mainstreaming action on climate change through participatory appraisal. *International Journal on Innovation and Sustainable Development, 1*(3), 238–259.

Young, O. (2008). Building regimes for socioecological systems: Institutional diagnosis. In O. R. Young, L. A. King, & H. Schroeder (Eds.), *Institutions and environmental change. Principal findings, applications, and research frontiers.* Cambridge: MIT Press.

The Value of Science and Technology Studies (STS) to Sustainability Research: A Critical Approach Toward Synthetic Biology Promises

Eleonore Pauwels

1 Introduction: Sustainability: A New "Venture" for Science and Technology Studies?

Sustainability has emerged as the newly ascendant policy issue of the twenty-first century. While we continue to argue about the true definition of "sustainability" – particularly since it has become a fashionable buzzword for the policy community and related funding agencies – the challenge of converting our present socio-technical system to a "sustainable" system has developed as a new master narrative, inspiring policy discourses both in Europe and the United States.

Sustainability science and policy are situated at the intersection of two transformations with in-depth ramifications as to how we conceive the world: one regarding the production and assessment of knowledge; the other about the very foundations of politics.

Issues of social and policy concern, like sustainable development, are conventionally assumed to be knowable through science, awaiting only "technical fixes." Yet, I would like to argue that the meaning and implications of sustainability as a policy issue are not intrinsic, but, for the most part, a human construction (Wynne 2007b). In the case of environmental governance, for example, measures for dealing with uncertainty and precaution, methods for storing and assessing data and, more generally, approaches to understanding the dynamics of the human-nature relationship, are not only structured and constrained by natural realities, but also socially and normatively shaped.

On the political front, the increasing focus on sustainability has largely changed the way we frame, conceive and discuss politics. According to Beck, "we require new, exploratory ideas and schemata, for example, 'reflexive governance', in order

E. Pauwels (✉)
Eleonore Pauwels, Research Scholar, Woodrow Wilson International Center for Scholars, One Woodrow Wilson Plaza, 1300 Pennsylvania Avenue NW, Washington D.C. 20004-3027 202/691/4359
e-mail: eleonore.pauwels@wilsoncenter.org

C.C. Jaeger et al. (eds.), *European Research on Sustainable Development*, DOI 10.1007/978-3-642-19202-9_9, © European Union, 2011
Published by Springer-Verlag Berlin Heidelberg 2011

to describe, understand, observe and explain the shifts now occurring in the very foundations of political action" (Beck 2006:31). Indeed, we are witnessing a progressive weakening of the authority of nation-states, coupled with disruptive global economic dynamics, which both require rethinking and re-organizing the space and contours of collective political action. This shift has diminished the connections between states and citizens, reducing the capacity of national governments to handle their citizens' expectations (Jasanoff 2005). While supranational concerns, such as the demand for sustainable development, are gaining political salience, policy leaders and officials fear that the necessary civic confidence may fail to transpire.

These are complex challenges emerging from powerful and pervasive socio-political forces. As such, I suggest that new avenues should be found to develop collective and critical approaches to understand the multiple meanings and normative dimensions of the notion of sustainability. There needs to be deliberate transdisciplinary and collective exploration of the socially and normatively constructed dimensions of sustainability, and in particular, to define the trajectories for research and innovation.

One of the ways to achieve this is with the development of inter and transdisciplinary research to understand the dynamics of human action, production of knowledge and environmental change. Encouraging collaboration across disciplines may lead to the creation of spaces, such as institutional forums and related mechanisms – where discussions about sustainability commitments can take place under the light of questions, uncertainties and ambiguities that are motivated by multiple disciplines.

This contribution explores the extent to which Science and Technology Studies (STS), rather than existing as a mere playing field between natural and social sciences, offers solid and productive theoretical models to approach sustainability challenges. For emerging technologies like synthetic biology that have been branded as "sustainable," Science and Technology Studies provides useful critical lenses to approach (1) technoscientific promises and their actual relevance to sustainable development; (2) socio-technical and socio-ecological alternatives in the development of these new technologies; and (3) the dynamics and interests at stake in the co-evolution of these technologies and society.

The purpose of this paper is to build a preliminary research agenda for sustainability grounded in the intuitions, lessons and theoretical insights derived from STS. To this end, the first section will discuss the value of STS to sustainability and initiate some reflections designed to help STS scholars better integrate sustainability into their research on synthetic biology. The second section will outline decisive research pathways which address the inherent ontological, epistemological and normative dimensions of sustainability and identify core perspectives brought by STS to the study of sustainable development. Inspired by the discussions of a group of experts in STS, sustainability science and synthetic biology,[1] I also consider a set of research practices and their related infrastructures

[1]On May 10 and 11, 2010, the Science and Technology Innovation Program at the Woodrow Wilson International Center for Scholars organized with the support of the University of Virginia and the U.S. National Science Foundation, a 2-day workshop to promote discussions between experts from STS, sustainability science and synthetic biology. This chapter is inspired by the discussions that took place on May 10 and 11, 2010.

that need to be built into the *Science and Society* agenda, as it addresses the challenges raised by sustainable development and emerging technologies.

2 Taking Constructivism Seriously

My primary question considers the value of Science and Technology Studies to sustainability. With this broad question in mind, the next section will begin on a wide canvas, with a review of the impact of recent work in Science and Technology Studies on the sustainability discourse – along the traditional notions of "technological progress."

2.1 *A Critical Approach Toward the Notions of Sustainability and Progress*

In the wake of the Enlightenment culture, notions of progress have unquestioningly been viewed in "western" governance discourse as the harbinger of better times. If society's view of scientific enquiry has become more sophisticated and nuanced, scientific and technological progress continues today to be considered worthy goals in and of themselves. Western nations have tied their visions of scientific research to that of economic competitiveness through continual technological innovation (Aho Report 2006; NAS 2009) (BLF, NRC Report 2009). A corollary is the production of normative discourses, in which parables of scientific or technological innovations are used to legitimate their inherent social value.

Against this background, leading STS academics suggest that dominant assumptions about science, sustainability and progress – which implicitly define existing institutional approaches to these issues – need to be fundamentally rethought (Wynne 2007b; Voß et al. 2006). They argue that "'scientific' object – sustainable development, 'safe limits' to human interference with climate, or 'risk', for example – is itself ambiguous, and in need of continual collective work to negotiate and at least temporarily stabilize its collective meaning" (Wynne 2007b:17).

Additionally, the concept of "sustainable science" conveys an array of complex, ambiguous, and discrepant positions over knowledge, values, meanings and interests that would lose from being reduced to questions of "technological fix." The related societal, ethical, and political controversies are only occasionally considered pertinent to "sustainability goals," raising the concern that other social dimensions are also being ignored. This reflects the fact that "sustainability" is not a "revealed" concept, but a contested one. Its substantial content cannot be analyzed exclusively through objective and factual scientific discourses. Instead, it will always include normative meanings that develop in the process of social interaction (Stirling 2006). Sustainability as a pathway to improve our social tolerance and the resilience of our systems of governance is an ambiguous and moving target. This is particularly true when it comes to questions of social need and prioritisation in defining the trajectories for scientific research and technological innovation.

Is Synthetic Biology the New Technological Fix?

Synthetic biology combines the principles of biology with the practices of computer engineering to build living machines from off-the-shelf chemical parts. Although synthetic biology is often confused with traditional genetics, since both seek to manipulate the building blocks of life, it nevertheless possesses a crucial difference: it seeks to *produce* genetic material from scratch, rather than modifying or copying material of existing organisms.

Narratives of technoscientific progress – such as those which combine general societal "progress" with technological "advance" – have existed for decades and, in this context, synthetic biology is not an exception. Synthetic biology, with its aim to engineer biological pathways, lies at the heart of what the U.S. National Research Council (NRC) has called *A New Biology for the 21st Century* (2009). This report recommends that a "New Biology" approach – one that depends on greater integration within biology, and closer collaboration with physical, computational, and earth scientists, mathematicians and engineers – be used to find solutions to four key societal needs: sustainable food production, ecosystem restoration, optimized biofuel production, and improvement in human health.

Synthetic biology is also presented in the U.S. press coverage as a key solution – a "technological fix" – to address the challenges of sustainable development by "greening" chemical and engineering sciences. As mentioned in the *San Jose Mercury News*, "Just as the first wave of biotechnology revolutionized agriculture and medicine, scientists today herald synthetic biology as a second wave of innovation capable of solving society's most pressing challenges. In the laboratory, researchers are developing customized organisms with powerful new capabilities. These modified cells can be programmed to fight diseases, create new wonder materials for manufacturing or produce an abundant source of clean, renewable energy."[2]

Synthetic biology in the scientific community discourse is thus often staged as the solution to a range of social ills, including the problematic sustainable development. However, opposite perspectives emerging from the civil society sector are voiced in the press to contest this: "Fearing that 'frankencells' will threaten the ecosystem, environmental groups such as Greenpeace and Friends of the Earth have labeled synthetic biology 'genetic engineering on steroids' and condemned it as 'a grave biosafety threat to people

(continued)

[2]D. Ballon, Opinion – "Synthetic Biology is a key to energy independence," *San Jose Mercury News*, 12/15/2008.

and the planet.'"[3] Some activists have already called for a complete research moratorium. Up to date, there is no solid reason to deny or question that synthetic biology may offer an unprecedented opportunity to transform modern medicine, generate clean biofuels and promote more sustainable infrastructures. However, several voices from the academic sector have warned that the technology may develop in an unsustainable way – in regard to environmental and societal concerns. In a report published in 2009, Michael Rodemeyer identified specific cases where research processes and infrastructures used to develop synthetic biology products of first, second and third generations will need more sophisticated risk assessment procedures than those on which the U.S. federal agencies currently rely (Rodemeyer 2009). In her testimony to the U.S. Presidential Bioethics Commission (July 8, 2010), Alison Snow systematically described how ecosystems might be impacted by the environmental release – intentional or unintentional – of synthetic organisms.[4] Current disagreements about the management of synthetic biology make it a particularly apposite lens through which to analyze the wider uncertainties about the relationship between science, society and sustainability.

2.2 Productive Theoretical Models: Portability and Co-production of Knowledge

As mentioned, decades of studying the social construction of science and technology have urged us to reconsider categorical assertions of objectivity and progress. Substantial time has passed since Mertonian norms were considered the "Holy Grail," defining "Science" as the institution capable of liberating the truth of nature from social and cultural horizons. More recent work in the social studies of science has, on the contrary, emphasized the portability of scientific knowledge. Among others, Bruno Latour, sociologist and philosopher of science, has relied on this concept to demonstrate the capacity of science to produce independent objects which exist "out there" in nature and are knowable exclusively through scientific methods (Latour 1987). The production of "scientific object" theory is based on an *in-principle* indisputable distinction between nature and culture that relegates the non-humans – supposed to exist independently of human agency – to the natural world. Central to the Science and Technology Studies enterprise has been the revival of these "hybrids" that traditional scientific views aimed to distinguish between the different spheres of nature and culture. Yet, the world around us is

[3]Idem.

[4]Pr. Alison Snow gave her testimony to the U.S. Presidential Commission for the Study of Bioethical Issues in the session entitled "Benefits and Risks" on July 8, 2010, at the Ritz-Carlton in Washington D.C. The testimony is available at: http://www.tvworldwide.com/events/bioethics/100708.

full of "hybrids" or "boundary objects," such as synthetic engineered microbes, genetically modified crops, acid rain, and climate change itself. Interestingly, the reason that these hybrid entities are bound to provoke endless disputes between scientists, environmentalists and other stakeholders, is because they inevitably depend on *co-produced* knowledge to be legitimated.

The theme of co-production – the simultaneous production of natural and social orders – has been salient in Science and Technology Studies for over a decade. "Through such investigations, it has been possible to demonstrate that the products of the sciences, both cognitive and material, embody beliefs not only about how the world is, but also how it *ought* to be. [...] Accordingly, to understand how social entities such as the "state" or natural entities such as the "gene" function in the world, one has to ask how diverse actors use and understand the concept, how it is articulated through formal and informal practices, where and by whom it is contested, and how it reasserts itself in the face of challenges to its integrity or meaning" (Jasanoff 2005: 19). The construction and stability of knowledge ultimately depends on the valuation of existing institutions, discourses, representations and common social practices. For example, the practices of environmental science are supported, even justified to some extent, by other social practices – including normative discourses.

The theoretical model of co-production that is developing out of Science and Technology Studies provides a window for analyzing the daunting array of questions, tensions, ambitions and concerns raised by sustainable development. Recognizing some of the connections between science and society in the making of knowledge may lead us to critically evaluate and question the construction of ethically and politically sustainable images of human-nature relationships.

Engineering Life or Engineering for "Better" Life?

By "engineering life," synthetic biology may have an unprecedented impact on the contemporary dynamics of human-nature relationships, with special attention on the beliefs and ideas that shape how people understand and value nature and assign it meaning in their lives. Synthetic biology may also have an effect on tightly coupled social and technological arrangements – what STS experts term *socio-technical systems* – that order human-nature relationships. In the long term, synthetic biology and its applications may bring about ontological changes and reclassifications in the world, producing new entities and new ways of understanding old ones. Such changes may entail a fundamental rethinking of the identity of the human self and its place in larger natural, social and political orders.

The following vision described by Rob Carlson is a good example of the potential changes we may be facing in the coming decades (Carlson 2001:1): "In 50 years, you may be reading *The Economist* on a leaf. The page will not look like a leaf, but it will be grown like a leaf. It will be designed for its function, and it will be alive. The leaf will be the product of intentional biological design and manufacturing. Rather than being constantly green, the

(continued)

cells on its surface will contain pigments controlled by the action of something akin to a nervous system. Like the skin of a cuttlefish, the cells will turn color to form words and images as directed by a connection to the internet of the day. Given the speed with which the cuttlefish changes its pigment these pages may not be fast enough to display moving images, but they will be fine for the written word. Each page will be slightly thicker than the paper *The Economist* is now printed on, providing room for control elements (the nervous system) and circulation of nutrients. When a page ages, or is damaged, it will be easily recycled. It will be fueled by sugar and light. Many of the artifacts produced in 50 years and used in daily living will have a similar appearance, and have similar origin. The consequences of mature biological design and manufacturing are widespread, and will affect all aspects of the economy including energy and resource usage, transportation, and labor."

2.3 *Transdisciplinarity and Collective Experimentation*

As introduced in the previous section, recent research in Science and Technology Studies has been conducted with the goal to re-define how scientific knowledge is produced and, subsequently, how this generation of knowledge will fit into the functions of society. Several STS academics have suggested that the traditional "Republic of Science" is being replaced by a new "Mode 2" of knowledge production (Gibbons et al. 1994). Two properties linked to this new "Mode" – transdisciplinarity and an orientation toward problem-solving – are particularly relevant for our discussion.

Analysis of long-term transformations in socio-ecological systems such as energy production and consumption, transportation, agriculture, resource extraction and manufacturing, requires understanding the systemic interconnections to which these problems refer. Yet, they deal with an array of heterogeneous elements, ranging from chemical pollution, to ecosystems, scientific studies, economic parameters, policy-making processes and cultural values and concerns. The traditional model of disciplinary science does not fully consider the interdependence of social, technological and ecological systems. Instead, it focuses on a very specific range of elements and interconnections. Outside of the laboratory, however, researchers have to face the constant interaction of scientific processes with the systemic embedding of cause and effect in which these processes operate. Regarding sustainability problems in particular, spill-over effects extend well beyond the scope of how they are defined by conventional disciplines.

Confirming the theoretical added-value of the model of co-production, I submit that the inherent limitations of one-dimensional perspectives also apply to scientific methods of knowledge generation. As such, the transdisciplinary model of knowledge production – which draws upon and integrates empirical and theoretical elements from a variety of fields – may help in analyzing the interactions between multiple sets of actors as they interact in real world entanglements. As argued by Voß and Kemp, "Considering the heterogeneity of the elements that play a part in sustainable development, effective problem treatment calls for the use of methods of integrated knowledge production that transcend the boundaries between disciplines and between science and society" (Voß et al. 2006: 10–11). They also insist on the benefit of integrating "the tacit knowledge of societal actors" – which is "generated in interactive settings in which knowledge is co-produced by scientists and actors from respective fields of social practice". Concretely, citizen and "concerned groups" get actively involved in the process of knowledge production, with the consequent result that some interactions between scientists and lay persons become permanent and build trust and mutual learning by working together in hybrid collectives.

A subsequent challenge lies in finding practical ways to integrate complex forms of "transdisciplinary" knowledge-making and assessment with more inclusive forms of stakeholder engagement and citizen deliberation. A corollary is to work with the potential of stakeholders and citizens to be independently knowledgeable agents. Each stakeholder is capable of their own reflective thinking about collective rationalities, knowledges and responsibilities. According to the 2007 Wynne report (p. 18), "this may lead to develop the cultural and political conditions under which widespread civic ownership of societal problems, like sustainability, and climate change (among others), and real engagement with the salient science, might be achieved".

In this respect, the European Commission is moving a step further. The recent internet public consultation on the "Nano Code of Conduct"[5] constitutes an effort not only to communicate science but also to set up a framework in the form of guidelines for carrying out responsible nanotechnology research. By doing so, the Commission wishes to internalize public consideration of potential implications of nanotechnology research in the research process itself and to participate in the ongoing global dialogue on the socio-economic impacts of nanotechnologies.

Besides the governance of nanotechnologies, the European Commission is also funding cooperative research processes between researchers and civil society organizations, which include both research and dialogue on relevant societal issues. There are other rich examples of newly emerging groups of actors in the field of science and innovation such as, the involvement of patient organisations and the elderly in the development of new technologies for health and social care. These citizen groups who engage with research constitute another layer of the system of knowledge production – often called the third sector of knowledge production – and bring a different logic into the knowledge-making process with the subsequent added-value of being capable of experimental practices and, ipso facto, of exploring alternatives to our socio-technical systems.

[5]Available at : http://europa.eu/sinapse/directaccess/science-and-society/public-debates/nano-recommendation/.

"Syn-Bio": A Complex and Transdisciplinary Science

Synthetic Biology is at the front edge of a wave that the US National Science Foundation (NSF) has termed "converging technologies" and involves bio, info, nano, and cognitive sciences. A lot of innovation will occur in the interstitial spaces between these disciplines, but the emerging multi-disciplinary smorgasbord will create challenges in terms of the ability of new fields to regulate their own actions, anticipate unintended consequences, communicate effectively with each other and the public and solve what some political scientists call "collective actions."

There will likely be new challenges in managing ethical, social and legal issues at the boundaries between disciplines. For instance, the emergence of biohacking reveals a growing culture of people interested in playing with genetic software and hardware in much the same way conventional hackers play with computer software and hardware. The key question then becomes: How do you establish a framework for socially and ethically responsible development of synthetic biology when the person you need to reach is an adolescent teenager constructing new biological code in his basement? Another potential concern may be that engineers entering the domain of biology have different ethical norms, standard practices and professional expectations vis-à-vis regulators and the public. Many engineers have little training in biology, toxicology, environmental sciences, and ecology – all of which are crucial for impact assessments of new biological organisms.

Synthetic biology thus crosses important technological frontiers, like the boundary between science and engineering, and is part of what has been called by the NRC the "Coming Biology Revolution." Such a revolution in the life sciences, its nature and goals, would ideally call for parallel transformations under the chapter of societal governance, but despite the efforts of visionary researchers to overcome the divisions between the two cultures of humanities and natural sciences (Jasanoff 2003), the new biology gets imagined mainly under the auspices of biologists, other natural scientists, mathematicians and engineers. A comprehensive understanding of the epistemic, ontological and normative changes induced by this new biology paradigm would require the involvement of researchers from humanities, social and environmental sciences, including STS and sustainability science.

2.4 *Political Conditions for Dissenting Imaginations*

Alongside the growing enthusiasm for the early participatory style of knowledge production, assessment and deliberation, several STS researchers have voiced their concerns that these emerging technologies of public elicitation might create a new form of technocracy by stripping away the political dimension of both science and participation (Wynne 2005, 2007a; Lezaun and Soneryd 2007). These concerns reflect recent STS interpretations of participatory approaches which lower democratic expectations but focus on understanding the conditions and perspectives under which these kinds of approaches might promote more plural and overtly political understandings of sustainability (Stirling 2008). Recent STS research has increasingly argued in favour of plural forms of engagement with civil society organisations and other groups of concerned citizens, which combine "governance from outside system" – such as upstream public engagement – and engagement "driving on inside of system"[6] – such as the above mentioned cooperative research processes (Stirling 2008, 2009; Fisher et al. 2006).

As we have argued under the co-production paradigm, knowledge is not simply transmitted from science to people, but is actively translated, construed and renewed in the context of emerging uncertainties, ambiguities, and collective experiences. On this basis, an array of STS scholarships has shown the decisive role played by civil society actors toward encouraging institutional reflections: to get decision-makers to question their own assumptions and consider a wider range of alternatives in face of these uncertainties and ambiguities (Wynne 2005, 2007a, b; Wilsdon 2007; Dryzek et al. 2009; Jasanoff 2003). According to the 2007 Wynne report (p. 78), "autonomous agents of civil society act and interact epistemically and socially in their own independent worlds and cultures. Against the dominant narrative of a singular hierarchy of knowledge, with publics imagined as epistemically-incompetent, thus untrustworthy, European institutions have the evidence-base to attribute a more active and creative role to their publics – and, as a result, to further encourage such a social capacity."

The evidence-base that has been taken into account by European policy-makers is the capacity exemplified by different layers of civil society actors in the GMs crops and food controversy to be attentive to what Wynne calls an "epistemic other" (2009: 13): "it is difference manifesting itself as an unknown set of realities, acting themselves as unknowns and beyond our control (but not beyond our responsibility), into a world we thought we controlled." On the surface of this epistemic variety, a democratically-committed knowledge-society is supposed to have the scientific and political imaginations to work out how a plurality of social actors could share knowledges, practices, and experiences with diverse scientific, policy and economic actors (Jasanoff 2009). Central to comparative studies within STS is the emphasis on "civic epistemologies" shaping the democratic practice of science and technology. Beyond the distributed nature of expertise, what matters most is the consideration of the divergent socio-cultural contexts in which technoscientific politics take place, *inter alia* the modes of knowledge-making in the public sphere and the levels of accountability and trust in the knowledges produced (Jasanoff 2005).

[6]Both expressions "governance from outside systems" and engagement "driving on inside system" have been eloquently described by Andy Stirling in the Session "Sustainability and Emerging Technologies" at the 2009 Conference of the Society for Social Studies of Science (4 S), October 29, 2009.

The above propositions progressively move us from "Mode 2" of knowledge production to the "Agora" – "where science and innovation interact with societies"[7] – and provide a role for public engagement of a more complex kind. In this case, scientists, engineers and policy-makers, sensitised through engagement to wider social imaginations, might decide for themselves to approach science and innovation differently. Interestingly, what lay expertise help picture, are hidden (1) *questions*, (2) *connections* and (3) *suggestions* (Wilsdon 2007): (a) interrogations might be about *what we don't know* and *how to learn what we should not do*; (b) connections might show the risky entanglements between public-private, global–local interests involved in techno-scientific promises; (c) suggestions might range from anecdotal evidence to alternative practice or alternative technology scenario. As explained by Stirling about current discourses on sustainability (Stirling 2009: 5):

> Often, the position is expressed as if there were 'no alternatives.' The questions asked are thus typically restricted to 'yes or no?'; 'how much?'; 'how fast?' and 'who leads?' If we move instead to more plural understandings of progress, then the quality of debate – and of the ensuing choices – thereby stands to be enriched. Instead of fixating on some contingently-privileged path, we might ask deeper, more balanced and searching questions about 'which way?'; 'what alternatives?'; 'who says?' and 'why?' This is the essence of a normative, analytic, epistemic, ontological – and consequently intrinsically political – project of 'pluralising progress'.

In a nutshell, the above-mentioned STS scholars eloquently demonstrated the importance of creating political conditions for dissenting imaginations (Wynne 2009; Jasanoff (2009). Beyond designing ideal participatory-governance processes, the prelude is intended to more reflexively understand the political background within which actors from different fields of social practices will be invited to interrogate particular framings of socio-technological regimes and their potential transition pathways, and to re-open them for debate (Stirling 2008; Smith and Stirling 2008). In this journey toward change-oriented research and policy, there is a necessary need for "daring to imagine" (Wynne 2009), for reflexivity and for empowerment as suggested by Jamison (2010: 13) "change-oriented research is about empowerment, by which the researcher applies knowledge gained from experience to processes of social learning, carried out together with those being 'studied'."

Jamison's reflection left us with a daunting yet challenging array of questions of how to promote empowerment, how to dare to imagine change and its uncertainties and how to open controversies to discussion. In this perspective, the work of Arie Rip brings remarkable insights into the value of (1) enabling future-oriented actions between actors who share an environment and (2) supporting them to create narratives about the potential resulting uncertainties and ambiguities (Rip 2006). More importantly, Rip stresses the need for *diversity* as a source of renewal by creating grey zones and interstitial spaces in existing orders and institutions where

[7]This concept of the "Agora" was introduced by Andy Stirling in the Session "Sustainability and Emerging Technologies" at the 2009 Conference of the Society for Social Studies of Science (4 S), October 29, 2009.

dissenting imaginations might be voiced. This reminds us that the ultimate challenge is about reviving atmospheres of our democracies which allow for the expression of dissenting imaginations. The ultimate challenge is avoiding "high-pace technoscientific politics" to withdraw from the democratic scene (*learning is forgetting*) and to cultivate the ability of "making things public," of turning "matters of facts" into "matters of concern" (Latour 2005). In a vibrant call, Latour invites us to give a chance to what he names *Dingpolitik* (2005: 23).

> The point of reviving this old etymology is that we don't assemble because we agree, look alike, feel good, are socially compatible or wish to fuse together but because we are brought by divisive matters of concern into some neutral, isolated place in order to come to some sort of provisional makeshift (dis)agreement. If the Ding designates both those who assemble because they are concerned as well as what causes their concerns and divisions, it should become the center of our attention: Back to Things!

Finally, at the core of this vibrant call for returning to *Dingpolitik* lies the diagnosis that the *modus vivendi* between modern democracies and technoscience has become increasingly compromised. The transformative power of technoscience reshapes societies in destabilizing ways, by imposing certain normativities and replacing controversies with "safe and serious" forms of knowledge which have significant ramifications as to how we conceive the world. STS research takes precedence over this diagnosis to investigate the conditions of emergence, both political and cultural, of collective practices for democratizing innovation – specifically when innovation aims at promoting sustainability. It encourages us to avoid substituting Enlightenment with "Sustainability normativity blinkers," but rather to focus on uninvited parties (*inter alia*, civil society actors, concerned group of citizens, and researchers favoring qualitative explorations of sustainability's meanings) as a "cauldron of concocting normativities" for emergent sustainability institutions.[8]

Throughout this theoretical section, I have tried to highlight the value of Science and Technology Studies for the research agenda on sustainability. In short, STS academics bring home the extent to which sustainability is raising long-term and collective issues that hinge on the political as well as the intellectual question of how much confidence we place in our knowledge (and in what forms of knowledge) which we use to legitimate new interventions in nature. The salient properties of our knowledge-making process are the following:

- In key aspects, such as the way their boundaries are framed, their dimensions selected and their meanings defined, issues of social and policy concerns – like sustainability – are partially human constructed.
- Changes in the modes of knowledge production have made science more embedded in society and more closely tied to its applications.

---

[8]Both expressions "Sustainability normativity blinkers" and "cauldron of concocting normativities" have been eloquently used by Andy Stirling in the Session "Sustainability and Emerging Technologies" at the 2009 Conference of the Society for Social Studies of Science (4 S), October 29, 2009.

- The renewed attention being directed towards transdisciplinarity and collective experimentation represents a promising development in defining trajectories for research and innovation.
- The transition from "Mode 2" of knowledge production to the *Agora* – where science and innovation interact with society – elaborates on the diagnosis of failure in developing open, inclusive, and diverse mechanisms of accountability in technology innovation. On this basis, this transition requires political conditions that promote plural and reflexive social normativities of progress.

Food for Thought: Make Synthetic Biology a "Matter of Concern," but How?

Below are a few questions about synthetic biology research developments that might help to produce the greater "reflexivity" of "science-in-the-making". At the very least, these questions might highlight some of the issues involved in synthetic biology research trajectories.

Emergence of Synthetic Biology Research Trajectories
To what extent do the synthetic biology research trajectories integrate the paradigm of sustainability? How are the meanings of "sustainability" negotiated in the rhetoric and economics of synthetic biology promises? To what extent do the synthetic biology practitioners themselves reflect about these trajectories – especially in terms of "sustainability transitions"? What ideas (concepts, beliefs, knowledge, ethics and values) underpin synthetic biology practitioners' understanding of nature, environment, science, technology and society as they relate to sustainability? What practices (behaviours, relationships, arrangements, and institutions) underpin the construction and maintenance of these ideas?

Actors, Dynamics and Configurations
Who are the leading actors (synthetic biology practitioners, the related stakeholders and celebratory institutions) and the marginalized actors ("non-invited parties") surrounding the development of synthetic biology? How do they present themselves through their research goals and practices? How do they define what would be a success in their research practices? To what extent is this definition of success effectively integrating sustainability research (and which exploration and meanings of sustainability)? What spaces are left for epistemic openings in an attempt to explore diverse meanings of "sustainability"?

(continued)

Related Social Science Researches and Their Practices of Engagement
To what extent do the lab-scale studies – which have flourished around the emergence of synthetic biology research[9] – lead to a better capacity to critically analyze synthetic biology promises and to collectively experiment with possible alternatives within synthetic biology? To what extent will they succeed in developing co-production among multiple disciplines and perspectives from the outset as opposed to downstream reflection upon the ethical, legal, and social implications of synthetic biology?

Entanglements Around the Notions of Ownership
What are the different models of ownership that are tacitly emerging inside and outside the laboratory, and within the public-private partnerships surrounding the development of synthetic biology? What reflections and analyses can be brought up by STS and Sustainability Science on this debate over different models of ownership? What are the implications of these different ownership models for our socio-technical systems, our socio-ecological systems, and our socio-economic systems?

3 Building a Science and Technology Studies Research Agenda for Sustainability

This section will outline possible research trajectories grounded in the work of STS scholars who focus on the social, philosophical and policy studies of science, technology and the environment.[10]

In this perspective, it is important to highlight some of the remarkable and exciting changes within recent STS research. STS research has become more relevant to understanding the co-production of science and technology with policy, democracy, law, and the management of environmental change, among other major institutional matters. Because of this, STS researchers have become increasingly involved with practices of technology development, policymaking, legal decision-making and governance in different fields, such as science & technology policy and environmental regulation.

[9]Two collaborative lab-scale projects might serve as field work: the *Human Practices* Laboratory directed by Paul Rabinow within the NSF-sponsored SynBERC project (http://www.synberc.org/content/articles/human-practices); and the Center for Synthetic Biology and Innovation as a collaboration between the BIOS Center (LSE) and the synthetic biology team of Imperial College (http://www.lse.ac.uk/collections/BIOS/synbio/synbio.htm).

[10]Sections II and III are based on exchanges and discussions in which I took part during the Workshop "Science, Technology and Sustainability," held at the National Science Foundation, September 8–9, 2008.

The balance between observation and participation seems to have shifted in these consequential practices of engagement. Such engagement is likely to have consequences for research methodologies, for researchers' obligations toward different publics, and for the kind of products STS-researchers deliver. In addition, like other aspects of science and technology, interventions by STS researchers are subject to contingencies and negotiations that can lead to unanticipated consequences.

3.1 Ontological, Epistemological and Normative Dilemma

At the end of the nineties, visionary minds from STS and environment studies started to take precedence over the diagnosis that science is not responding adequately to the challenges of our times, and particularly, those posed by the quest for sustainable development. Recognizing the need for a new "Social Contract for Science," they essentially identified three types of challenges that future societies would have to cope with (Gallopin et al. 2001).

As a first diagnosis, society is facing ***ontological*** changes which encompass changes induced and driven by human behaviors impacting nature. These changes are proceeding at unprecedented rates and scales, and subsequently result in growing interconnections at many levels. These ontological changes are made visible and understood through the analysis of what STS research, among other fields, calls socio-technical systems, which consist of our large-scale technological infrastructures (such as transportation systems and energy distribution grids) embedded in a dense web of human and social values, behaviors, relationships, and institutions (Smith and Stirling 2008). Socio-technical developments, and the powerful applications created in their wake, intertwine science, technology and society dimensions – making understandings of both the human and social aspects of science and technology critical to analyzing and responding to sustainability challenges (Miller et al. 2008).

These ontological changes have progressively rippled our systems of knowledge production and assessment, changing the ways we view the world and inviting us to think in terms of connectedness, relationships, context and socio-ecological patterns (Gallopin et al. 2001). They have bound us to new ***epistemological*** challenges which would benefit from being addressed through STS lenses. Indeed, central to the STS endeavor has been to explore the human and social practices, as well as philosophical and ethical frameworks, that have determined how we have come to learn about and value aspects of sustainability research, such as society and the environment (Jamison 2001; Jasanoff 2005; Miller 2005; Norton 2005). STS scholarship has similarly developed a comprehensive expertise into future-oriented analyses facing the uncertainties, ambiguities and unpredictability that are built in the fabric of reality (Stirling 2006; Guston et al. 2002; Sarewitz 2005).

Building on this increasing expertise in anticipatory thinking, researchers have also developed experimental modes of participative foresight for science and technology governance as well as environmental and sustainability studies (Kasemir et al. 2003). This is only the tip of the iceberg as STS researchers have produced, among their core conceptual areas, critical analyses of the strategies used by political and policy institutions for governing science and technology in society. An array of STS researchers have been increasingly concerned with the politics of environmental sciences, articulating positions with respect to questions about the role of expertise in democracy (Miller and Edwards 2001; Jasanoff and Martello 2004; Irwin 1995; Fischer 2000), or engaging in studies that directly refer to questions of reform and activism (Smith 2005). This indicates why STS scholars are particularly entrenched in addressing the third kind of challenges – the ***normative*** dilemma that emerges in science and technology policy, management, and regulation.

The next sub-sections synthesize some key research questions which address the above-described ontological, epistemological and normative dimensions and might be identified as core perspectives brought by STS to the study of sustainability.[11]

3.2 Participatory and Anticipatory Thinking for Sustainable Socio-Technical Systems

Significant attention in STS has been paid to co-production within the functioning of socio-technological systems and socio-ecological systems (*inter alia*, Jasanoff 2004, 2005). This body of knowledge might be decisive to understand why and how these systems' structure and dynamics contribute to sustainable and unsustainable outcomes as well as why these systems have been designed that way and how they have become incumbent parts of our socio-technological landscapes. Building on these findings, STS research could help better understand the social dynamics that prelude to replacing incumbent socio-technical systems with alternative, more sustainable, ones.

In a nutshell, STS scholarship brings home critical insights into how human societies make choices impacting the design of current socio-technical systems and how these choices and their spill-over effects influence how societies envision the systems of tomorrow (Miller et al. 2008). As the discipline has developed a useful set of tools to analyze technology within society, it might be able to identify the social conditions that inhibit modern societies from choosing sustainable technologies and practices. For instance, STS researchers have started to focus on sustainability technologies as practices, including the understandings, learning and stabilization processes that underlie these practices. Sustainability practices might be explored through ethnographic studies as a complement to participatory settings

[11]Researchers in sustainability science have identified other challenges which pertain to public perceptions, such as cultural practices and social learning (Pahl-Wostl et al. 2008).

designed to understand conceptions of "current" practices, of their amenability to change, and of the ways in which they can be shaped into sustainable ones.[12]

As a parallel area of inquiry, STS research could significantly address the question of how sustainability is apprehended within the functioning of socio-technical systems with a subsequent focus on how these systems cope with their inherent ambiguities, vulnerabilities and collective responsibility. When it comes to sustainability, only limited research effort has been devoted to understanding the socio-economic impacts resulting from the introduction of new technologies within society. In contrast, considerable more attention has been paid to their impacts on the environment. But, here, hybridization in the process of assessing ecological and socio-economic implications would benefit from being explored. What choice of indicators will make it possible to track important social values within a particular socio-technical system? What are the social dimensions to be sustained? What tradeoffs have been made in the past between social and ecological sustainability and why? How can we better understand the dynamics within the social formations driving and hindering sustainability transitions?

Given that we live in social systems which are organized, for the most part, around a plurality of values, research aimed at exploring social meanings of sustainability should involve anticipatory and participatory thinking (Stirling 2006, 2009). They should also rely on empirical methodologies that are capable of guiding public deliberations toward visualizations and framings that endorse multiple and varied values (Shaw et al. 2009). STS researchers could reflect on models for engaging civil society actors and wider publics in processes of envisioning and assessing technological futures (Fischer 2000; Irwin 1995). Cooperative research processes between STS scholars, experts in engineering and sustainability science, and civil society actors could lead to create deliberative spaces where communities interact in practice and, *ipso facto*, contribute to the reflection on more sustainable socio-technical systems for the future.

3.3 Critical Assessment of Sustainability Knowledge Production and Valuation

As shown in our theoretical analysis, STS scholars have begun to demonstrate the importance of introducing exogenous normativity into discourses of progress and the role of marginalized and unconventional actors to play in the directions of innovation. Similarly, STS research can investigate the conceptual and epistemological premises of sustainability, and the social practices on which they are built.

[12]This is the approach applied within the EPSRC-supported project "CHARM" which includes research on electricity consumption. CHARM is coordinated by R. Rettie and K. Burchell, both at Kingston University. See: http://business.kingston.ac.uk/charm.

This investigation might help to deconstruct the values, reasoning and framings at stake in controversies over sustainability.

A related area in need of investigation includes empirical studies of the knowledge production and assessment practices not only within scientific disciplines but also within government agencies, corporations, and other social formations concerned with the environment. Indeed, when it comes to sustainability research and decision-making, there is a need for empirically-based analyses of how knowledge systems work and how they might become better integrated with decision-making. Such empirical studies might reveal the social and political arrangements that prevail in knowledge and production assessment, the opportunities for opening these processes up to alternatives, and the pathologies of closing up (Stirling 2006). For instance, a growing body of STS research has begun to explore the various ways that "green knowledge" is being constituted and mobilized (Jamison 2001). How is sustainability expertise produced, distributed, and, subsequently, transformed through decision-makers' interpretative frameworks and political agendas? How are local knowledges, skills and technologies evaluated and mobilized towards sustainability? What is the role of social formations (social movements, business and government actors) in producing knowledges and arbitrating environmental controversies?

Such controversies have often been used by STS scholars as a basis for investigation. Further research might examine why there is a perceptible lack of controversy lying beneath areas that could become a "sustainability matter of concern." In this perspective, STS research is challenging the assumption in sustainability research – and potentially, in sustainability science – that there is already an agreement on the meanings of sustainability and on its social translations, measurement and realizations (Stirling 2006, 2009; Voβ and Kemp 2006; Norton 2005). A robust STS-centered engagement would significantly broaden current thinking in sustainability research by offering critical insights into how sustainability knowledge systems function in societies – for example, what has been exposed through "lab studies" – what their implications for community decision-making are, and how to confront them with processes of transition (Cash et al. 2003). The specific emphasis of the field on material practices, including human and non-human entanglements within knowledge production systems, constitutes a distinctive added value to other approaches to sustainability research (Jasanoff 2005; Latour 2004). More recently, research at the crossroad between STS and political ecology has explored case-studies that connect dynamics of local expertise with political and policy practices of expertise at national and international scales (Jasanoff and Martello 2004; Miller 2005; Miller and Edwards 2001). These studies might help reflect on improved models for transcending and connecting data generated in local and context specific sites to trends and challenges at the global scales.

Finally, when it comes to a "sustainability matter of concern," the question of our collective ignorance might be as interesting as the question of our systems of production of knowledge (Proctor and Schiebinger 2008). What are the epistemic, social and political rationales behind our socio-ecological and socio-technical ignorance? There might be some room here for STS scholars to problematize the sociological and historical roots of the dynamics that lead to non production of knowledge about what and who we are supposed to "sustain".[13]

3.4 Democratizing Sustainability Innovation: Deliberating about Socio-Technological Futures

The above argument reminds us that at the core of the deliberations regarding our socio-technological futures lie collective questions about the nature and scope of what we want to sustain. These questions involve normative positions about human-nature relationships with a particular focus on the values that influence how societies understand nature and assign it meaning in their lives (Norton 2005). Deliberation about these collective preferences imply being able to track democratic tradeoffs and contradictions within sustainability issues and, subsequently, better integrating normative aspects into decision-making (Cash et al. 2003). This is particularly crucial when sustainability challenges give rise to necessary and large-scale changes. How can change be promoted in our current socio-technical systems and whose agencies and responsibilities are at stake? How can social practices of human decision-making be conducted to improve the aptitude to design more sustainable socio-technical systems in the future?

Because it gives special attention to the interactions between epistemic and political processes, STS can genuinely study knowledge and technology as integral parts of policy-making. Therefore, it has been able to reflect on and design more reflexive and inclusive approaches to governance of science and technology (Stirling 2009; Smith and Stirling 2008). This expertise might be extremely valuable when starting to think about comparable models of policy and decision-making oriented toward the transformation of incumbent socio-technical systems. Further in-depth expertise will be needed in the social practices of policy and decision-making related to scientific and technological change as well as collective experimentation to test inclusive practices of engagement with technical, business, policy and civic communities (Wynne 2007b).

On the theoretical front, it requires *inter alia* developing conceptual and methodological frameworks to tackle the inherent complexity and diversity of knowledges blended into sustainability decision-making processes (Grin 2006). This implies finding experimental ways to integrate complex forms of transdisciplinary assessment with more inclusive forms of stakeholders' engagement and citizen deliberation. Among these challenges, one is about improving the visualization of the socio-technological choices we are faced with (Shaw et al. 2009); the other is to foster networks that bring practitioners together with scholars to promote the co-evolution of diverse forms of knowledges (Rip 2006; Stirling 2005).

[13]The question "What is it we want to sustain?" was eloquently posed by S. Jasanoff at the Workshop "Science, Technology and Sustainability," held at the National Science Foundation, September 8–9, 2008.

This co-evolution could take the form of collaborative research networks that can collectively pursue the knowledges' synthesis, and cooperate with colleagues in technical, civic, entrepreneurial, and policy communities to translate research into new approaches to meet the sustainability challenges.

Finally, democratizing sustainability innovation entails rethinking the distribution of responsibilities within our complex socio-technological systems. Such rethinking must be an interdisciplinary and an inter-cultural process, in order to conform to important notions of democracy and justice. This reflection takes us back to our starting point. It finds its inspiration in a critic of the conventional linear Enlightenment understanding of scientific and technological progress viewed as endogenously determined, teleological and likely to impose self-referential normativity (Stirling 2009). It finds its inspiration in an array of scholarly work which has attempted to show why scientific, social and political actors have mutually failed to distinguish, as a "matter of concern," several roles and responsibilities of science in our globalized knowledge-societies (Wynne 2009; Jasanoff 2009).

4 Empirical Reflections at the Crossroad Between Science, Technology and Sustainability

I conclude this kaleidoscope of possible research trajectories, with empirical reflections which arise from the discussions of a group of experts in STS, sustainability science and synthetic biology that gathered on May 10–11, 2010, at the Woodrow Wilson International Center for Scholars, with the support of the U.S. National Science Foundation. The workshop led to intense cross-field reflections, debates and controversies on production of knowledge, impact on policy-making, and cross-national differences in the way research cultures reproduce and emerging technologies – like synthetic biology – interact with societies. It finally shed light on potential collaborations as well as research, education and policy initiatives at the crossroad between science, technology and sustainability.

I do not assume that these reflections are final and comprehensive, but hope that they can play a valuable role in stimulating further thinking and proposals for additional and consequent cross-field collaborations. They focus on the kinds of research practices and infrastructures necessary to make possible not only the pursuit of the research agenda outlined in Sect. 2, but also the capacity to translate its findings into concrete action to enhance sustainability in human societies. Key aspects of and questions concerning these research practices and infrastructure include:

- The development of collaborative research groups that can collectively pursue this research (cf. Sect. 2), combine its findings, and cooperate with colleagues in technical, civic, entrepreneurial, and policy communities to translate research into new approaches to meeting the sustainability challenge; The concept of "collaboration" itself provoked interrogations among the participants: How do you create the infrastructures so that complex ways of thinking from different fields can seat somewhere and learn from each other? How can we think about forms of "cohabitation" where researchers from different fields would reflect together on design, options, research questions and trajectories? Is there a possibility for different socio-technical imaginations to cohabit? What are the necessary conditions (institutional, epistemic, political and cultural) to develop different forms and places for *reflexivity*, at different levels, in different contexts and networks such as the educational systems, the policy systems or the laboratories?

- The creation of novel training programs that are able to prepare the next generation of researchers in the new methods and theories that emerge from the research agenda outlined in Sect. 2; transdisciplinarity appeared as one of the key features on which novel training programs should be built. Additionally, the concept generated new questions: What are the barriers to developing a transdisciplinary research program within universities or research centers that foster the type of partnerships needed in the assessment and governance of emerging technologies like synthetic biology (What are the impacts of cost structure, pressure from departments and power structure within universities)? How should we re-think the roles, goals and practices of knowledge-producers like universities, academies, and research centers when it comes to cross-field collaborations, especially with the aim of transitioning towards more sustainable socio-technical and socio-ecological systems?
- The fostering of networks that bring practitioners, policy-makers and scholars together to promote the co-evolution of diverse forms of knowledge; The notions of "impact," "intervention" and "channels of action from academia to policy-making" were explored in terms of opportunities for (1) theorizing transdisciplinary and systemic ways of critically assessing problems and producing knowledge about them, and (2) institutionalizing cross-field experiments: How can channels of influence on policy-making be maximized through cross-field collaboration? What are the obstacles? How can we build on funding schemes, publications, and public infrastructures to promote cross-field collaborations?

The following diagram builds on the above reflections to propose potential parallel, yet distinct, discussions for sustainability and synthetic biology (Fig. 1).

5 Conclusion: From Technoscientific Hubris to Socio-Technical Hybrids[14]

In the first section of this contribution, I highlighted the theoretical value of Science and Technology Studies to sustainability research. In this respect, an important conclusion of this contribution is that steps should be taken away from narrow and

[14]See the book by M. Hard and A. Jamison, *Hubris and Hybrids: A cultural History of Technology and Science*, New York and London: Routledge, 2005.

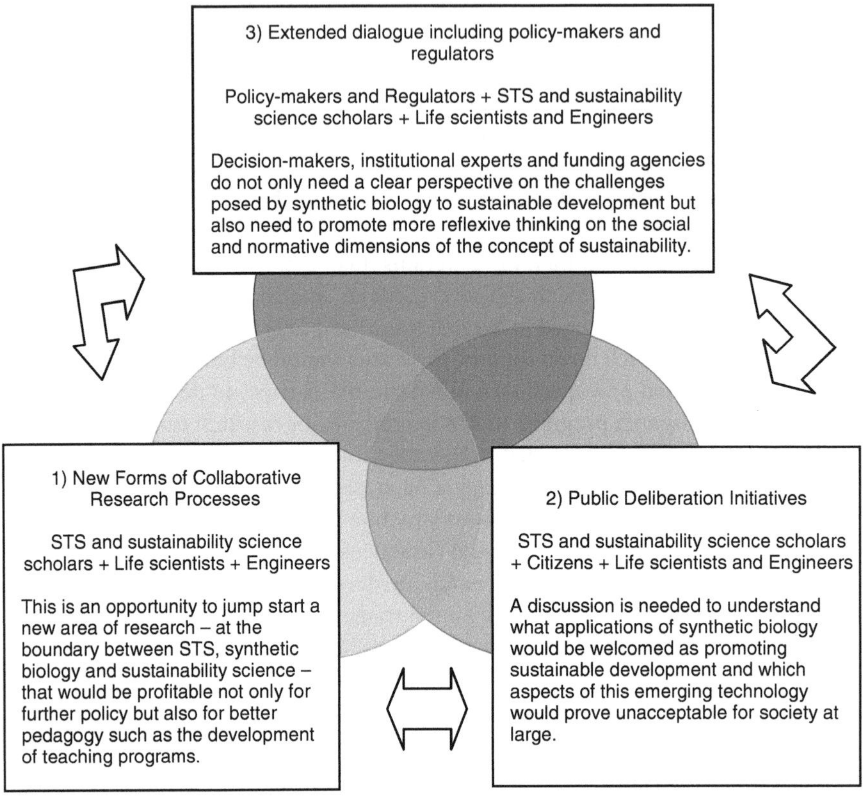

Fig. 1 Interactive representation of potential parallel but distinct discussions on synthetic biology and sustainability policy

exclusive understandings of knowledge production and assessment toward recognising more socially distributed autonomous and diverse collective forms of producing knowledge.

Recent discussions of Mode-2 science have pointed out that ways of producing techno-scientific knowledge already extend well beyond the classical mode of basic science. A stronger role of application contexts in the very production of knowledge, transdisciplinarity and preliminary attempts to develop collective experimentation and spaces for dissenting imaginations are but a few elements which indicate much broader social involvement in how knowledge is produced, contested and validated.

However, the public and policy debates surrounding synthetic biology have been narrowly focused around a utilitarian calculation of its technological benefits versus its potential regulatory risks. Although the technical aspects of synthetic biology policy are immensely important, spanning from controversies on ownership, socio-technical implications to biosecurity and biosafety concerns (nobody would like to the re-engineered flu virus mysteriously escaping from the lab), they ignore fundamental questions about *what* applications of synthetic biology should be considered sustainable, and thus limit the discussion to the opinions of a few technocratic elites.

Indeed, there are serious social, ethical and safety questions surrounding synthetic biology. The purpose of these questions is not to cause undue alarm, or advocate a knee-jerk form of neo-luddism. Rather, it is to expand awareness on what effects synthetic biology could have on both our political systems and our conception of humanity as a whole.

Ultimately, the research priorities and infrastructures identified in Sects. 2 and 3 acknowledge and focus on the inextricable human dimension of our socio-technical system; they also build on the social and collective practices and dimensions that characterize our forms of knowledge production and assessment, and related forms of decision-making.

If, in the real world, scientific and technological hubris encounter the wider societal context of values and aspirations, giving birth to novel constructions of technological artefacts and socio-organisational innovations, the case of sustainability might be a good example of such long "hybridisation" processes.

References

Aho Report, Aho, E., Cornu, J., Georghiou, L., & Subirá, A. (2006). *Creating an innovative Europe: European commission report of the independent expert group on R&D and innovation* (EUR 22005). Luxembourg: European Commission.

Bauknecht, D., & Kemp, R. (2006). *Reflexive governance for sustainable development* (pp. 31–56). Cheltenham: Edward Elgar Publishing.

Beck, U. (2006). Reflexive Governance: Politics in the Global Risk Society. In J.-P Voß, D. Bauknecht, & R. Kemp (Eds.), *Reflexive Governance for Sustainable Development (31–56)*. Cheltenham: Edward Elgar Publishing.

Board on Life Sciences: Division on Earth and Life Studies. (2009). *A new biology for the 21st century*. Washington: The National Academies Press.

Carlson, R. (2001). Open source biology and its impact on industry. *IEE Spectrum, 38*(5), 15–17.

Cash, D. W., Clark, W. C., Alcock, F., Dickson, N. M., Eckley, N., Guston, D., et al. (2003). Knowledge systems for sustainable development. *PNAS, 100*(14), 8086–8091.

Dryzek, J. S., Goodin, R. E., Tucker, A., & Reber, B. (2009). Promethean elites encounter precautionary publics: The case of GM foods. *Science, Technology and Human Values, 34*(3), 263–288.

Fischer, F. (2000). *Citizens, experts and the environment: The politics of local knowledge*. Chapel Hill: Duke University Press.

Fisher, E., Mahajan, R. L., & Mitcham, C. (2006). Midstream modulation of technology: Governance from within. *Bulletin of Science, Technology and Society, 26*(6), 485–496.

Gallopin, G. C., Funtowicz, S., O'Connor, M., & Ravetz, J. (2001). Science for the 21st century: From social contract to the scientific core. *International Journal of Social Science, 168*, 239–250.

Gibbons, M., Nowotny, H., Limoges, C., Schwartzman, S., Scott, P., & Trow, M. (1994). *The new production of knowledge: The dynamics of science and research in contemporary society*. London: Sage.

Grin, J. (2006). Reflexive modernization as a governance issue, or: Designing and shaping re-structuration. In J.-P. Voß, D. Bauknecht, & R. Kemp (Eds.), *Reflexive governance for sustainable development* (pp. 57–81)). Cheltenham: Edward Elgar Publishing.

Guston, D., David, H., & Sarewitz, D. (2002). Real-time technology assessment. *Technology in Society, 24*(1–2), 93–109.

Irwin, A. (1995). *Citizen science: A study of people, expertise and sustainable development.* New York: Routledge Press.

Jamison, A. (2001). *The making of green knowledge – environmental politics and cultural transformation.* Cambridge: Cambridge University Press.

Jamison, A. (2010). In search of green knowledge: A cognitive approach to sustainable development. In S. Moore (Ed.), *Pragmatic sustainability: theoretical and practical tools* (pp. 68–80). New York: Routledge. Retrieved from http://people.plan.aau.dk/~andy/In%20Search%20of%20Green%20Knowledge.doc

Jasanoff, S. (2003). Technologies of humility: Citizen participation in governing science. *Minerva, 41*(3), 223–244.

Jasanoff, S. (2004). *States of knowledge: The co-production of science and social order.* London: Routledge Press.

Jasanoff, S. (2005). *Designs on nature: Science and democracy in Europe and the United States.* Princeton: Princeton University Press.

Jasanoff, S. (2009, May). *Governing innovation.* Paper presented at the Symposium *Knowledge in Question* – a symposium on interrogating knowledge and questioning science # 597. Retrieved from http://www.india-seminar.com/2009/597.htm. Accessed Date: July 1, 2010.

Jasanoff, S., & Martello, M. L. (2004). *Earthly politics: Local and global in environmental governance.* Cambridge: MIT Press.

Kasemir, B., Jäger, J., Jaeger, C. C., & Gardner, M. T. (2003). *Public participation in sustainability science: A handbook.* Cambridge: Cambridge University Press.

Latour, B. (1987). *Science in action. how to follow scientists and engineers in society.* Milton Keynes: Open University Press.

Latour, B. (2004). *Politics of nature: How to bring the sciences into democracy.* Cambridge: Harvard University Press.

Latour, B. (2005). From realpolitik to dingpolitik or how to make things public. In B. Latour & P. Weibel (Eds.), *Making things public – atmospheres of democracy* (pp. 14–43). Germany: ZKM: MIT Press and Karlsruhe.

Lezaun, J., & Soneryd, L. (2007). Consulting citizens: Technologies of elicitation and the mobility of publics. *Public Understanding of Science, 16*(3), 279–297.

Miller, C. (2005). New civic epistemologies of quantification: Making sense of local and global indicators of sustainability. *Science, Technology and Human Values, 16*(4), 478–500.

Miller, C., & Edwards, P. N. (Eds.). (2001). *Changing the atmosphere: Expert knowledge and environmental governance.* Cambridge: MIT Press.

Miller, C., Sarewitz, D., & Light, A. (Eds.). Report of a workshop at the national science foundation on Science, technology, and sustainability: Building a research agenda. Washington, DC: National Science Foundation. Retrieved September 8–9, 2008, from http://www.nsf.gov/sbe/ses/sts/Science_Technology_and_Sustainability_Workshop_Rpt.pdf.

Norton, B. (2005). *Sustainability: A philosophy of adaptive ecosystem management.* Chicago: University of Chicago Press.

Pahl-Wostl, C., Tàbara, J. D., Bouwen, R., Craps, M., Dewulf, A., Mostert, E., et al. (2008). The importance of social learning and culture for sustainable resources management. *Ecological Economics, 64*(3), 484–495.

Proctor, R. N., & Schiebinger, L. (Eds.). (2008). *Agnotology: The making and unmaking of ignorance.* Palo Alto: Stanford University Press.

Rip, A. (2006). A co-evolutionary approach to reflexive governance – and its ironies. In J.-P. Voß, D. Bauknecht, & R. Kemp (Eds.), *Reflexive governance for sustainable development* (pp. 82–101). Cheltenham: Edward Elgar Publishing.

Rodemeyer, M. (2009). New life, old bottles: Regulating first-generation products of synthetic biology. *Synbio 2*. Washington, DC: Woodrow Wilson International Center for Scholars.

Sarewitz, D. (2005). This won't hurt a bit: Assessing and governing rapidly advancing technologies in a democracy. In M. Rodemeyer, D. Sarewitz, & J. Wilsdon (Eds.), *The future of technology assessment*. Washington, DC: Woodrow Wilson International Center for Scholars.

Shaw, A., Sheppard, S., Burch, S., Flanders, D., Wiek, A., Carmichael, J., et al. (2009). Making local futures tangible – synthesizing, downscaling, and visualizing climate change scenarios for participatory capacity building. *Global Environmental Change, 19*(4), 447–463.

Smith, A. (2005). The alternative technology movement: An analysis of its framing and negotiation of technology development. *Human Ecology Review, 12*(2), 106–119.

Smith, A., & Stirling, A. (2008). *Socio-ecological resilience and socio-technical transitions: Critical issues for sustainability governance* (STEPS Working Paper 8). Brighton: STEPS Centre.

Stirling, A. (2005). Opening up or closing down: Analysis, participation and power in the social appraisal of technology. In M. Leach, I. Scoones, & B. Wynne (Eds.), *Science and citizens: Globalization and the challenge of engagement* (pp. 218–231). London: Zed.

Stirling, A. (2006). Precaution, foresight and sustainability: Reflection and reflexivity in the governance of science and technology. In J.-P. Voß, D. Bauknecht, & R. Kemp (Eds.), *Reflexive governance for sustainable development* (pp. 225–272). Cheltenham: Edward Elgar Publishing.

Stirling, A. (2008). 'Opening up' and 'closing down': Power, participation, and pluralism in the social appraisal of technology. *Science, Technology and Human Values, 33*(2), 262–294.

Stirling, A. (2009). *Direction, distribution and diversity! Pluralising progress in innovation, sustainability and development* (STEPS Working Paper 32). Brighton: STEPS Centre.

Voβ, J.-P., & Kemp, R. (2006). Sustainability and reflexive governance: Introduction. In J.-P. Voß, D. Bauknecht, & R. Kemp (Eds.), *Reflexive governance for sustainable development* (pp. 3–28). Cheltenham: Edward Elgar Publishing.

Wilsdon, J. (2007, October 8–10). *Public engagement in science. Report of the Science in Society Session, Portuguese Presidency Conference – The Future of Science and Technology in Europe, Lisbon* (EUR 23334). Luxembourg: European Commission.

Wynne, B. (2005). Risk as globalizing "democratic" discourse? framing subjects and citizens. In M. Leach, I. Scoones, & B. Wynne (Eds.), *Science and citizens: Globalization and the challenge of engagement* (pp. 66–82). London: Zed Books.

Wynne, B. (2007). Public participation in science and technology: Performing and obscuring a political-conceptual category mistake. *East Asian Science, Technology and Society, 1*(1), 99–110.

Wynne, B. (2009). *Daring to imagine*. Paper presented at the symposium knowledge in question – A symposium on interrogating knowledge and questioning science # 597. Retrieved from http://www.india-seminar.com/2009/597/597_brian_wynne.htm. Accessed Date: July 1, 2010

Wynne, B., Callon, M., Eduarda Gonçalves, M., Jasanoff, S., Jepsen, M., Joly, P.B., Konopasek, Z., May, S., Neubauer, C., Rip, A., Siune, K., Stirling, A., & Tallacchini, M. (2007b). *Taking European knowledge society seriously: European Commission Report of the Independent Expert Group on Science and Governance* (EUR 22700). Luxembourg: European Commission.

Sustainable Development: Responding to the Research Challenge in the Land of the Long White Cloud, Aotearoa New Zealand

Richard F.S. Gordon

1 Introduction (May 2009)

Amongst the challenges facing us in New Zealand, three questions are relevant to the theme of this conference. First, what is the relevance of a Sustainable Development research agenda to an island nation of 4 million people in the grip of a global economic crisis? Second, how may our precious investment in research, science and technology be guided so as to maximise the return to the nation? Third, what are priorities for investment in sustainable development research? This paper explores some possible answers. These answers are inter-linked and they reflect several realities, for example: that research in New Zealand is a tiny proportion of the global whole, but New Zealand can be a laboratory for the world; that our research resources are limited, so what we do must have impact; that achieving impact in complex systems comes from influencing paradigms and mechanisms of governance[1]; and that different peoples have different world views and approaches to governance.

[1]Based on the higher leverage points (1–3) in Meadows D. (1999). Leverage points: places to intervene in a system. Sustainability Institute, Hartland Vermont. Available at: http://www.sustainer.org/pubs/Leverage_Points.pdf.

R.F.S. Gordon (✉)
Landcare Research Manaaki Whenua, P.O. Box 40, Lincoln, Canterbury 7640, New Zealand
e-mail: gordonr@landcareresearch.co.nz

C.C. Jaeger et al. (eds.), *European Research on Sustainable Development*,
DOI 10.1007/978-3-642-19202-9_10, © European Union, 2011
Published by Springer-Verlag Berlin Heidelberg 2011

2 What Is the Relevance of a Sustainable Development Research Agenda to an Island Nation of Four Million People in the Face of a Global Economic Crisis?

In 2003 the New Zealand government issued a Programme of Action for Sustainable Development.[2] This broke new ground in our country by identifying the changes in the way we do things, and specifically in the way government acts, that will be needed to make a success of sustainable development. It described a new way of thinking and working: looking after people; taking the long-term view; taking account of the social, economic, environmental and cultural effects of our decisions; and encouraging participation and partnerships.

In 2007 the then Prime Minister, Helen Clark, announced an intention to make New Zealand truly sustainable. She defined the sustainability challenge as "one of the defining global issues of the twenty-first century", and "a challenge that New Zealand must meet to protect our nation's unique way of life and our future prosperity." She talked of the need to share responsibility in this challenge.[3]

In 2009 New Zealand faces a similar challenge to other countries. The result of unsustainable financial practices at home and in the global community leaves us facing an economic hardship that is difficult to predict. We face a harsh reality that unsustainable behaviour is just that: *unsustainable*. The economic turmoil is a taster for the turmoil predicted as a result of unsustainable management of our environmental resources and global climate. Whether the defining issue will be climate or water, soil erosion, or loss of biodiversity, we face an uncertain but almost certainly punishing future.

The economic crisis may support the old adage, "*it is hard to be green when you are in the red*". Some think we may literally be unable to afford environmental measures in the short-term that are necessary for long-term welfare. Therefore it is encouraging that many national recovery packages appear to include environmental initiatives, for instance in clean technology.[4] But we will miss a significant lesson if we do not recognise that in addressing the economic crisis we may develop some of the tools we need to address a potentially greater environmental and social crisis looming in the next few decades as a result of climate change and the depletion of natural capital. It may help us to shift paradigms and improve governance systems for lasting benefit to society.

---

[2]Department of Prime Minister and Cabinet (2003). Sustainable development for New Zealand: programme of action. Available at: http://www.mfe.govt.nz/publications/sus-dev/sus-dev-programme-of-action-jan03.html.

[3]Helen Clark in Voices for Sustainability. Available at: http://www.landcareresearch.co.nz/sustainability/sustainabilty_details.asp?Sustainability_ID=59.

[4]For example, Green technologies win UKP1.4billion in UK budget in Nature. 22 Apr 2009. Available at: http://www.nature.com/news/2009/090422/full/news.2009.392.html.

Returning to our initial questions: What is the relevance of a sustainable development research agenda in the face of an economic crisis? Scientists might say the crisis is an experiment in how society makes the transition from an unsustainable to a sustainable system. What is the special relevance to an island nation of 4 million people? Scientists might say we have in Aotearoa New Zealand a useful laboratory, with clearly defined boundaries, reasonably well regulated internal conditions, fairly clear external influences, and a national characteristic attitude of "give it a fair go", meaning that we are pragmatic and willing to try new ideas. In this laboratory we may evaluate solutions of relevance to other countries.

3 How May Our Investment in Science and Technology Be Guided So as to Maximise the Return to the Nation?

One aspect of New Zealand pragmatism is evident in its approach to science funding. We conduct a tiny proportion of the world's science and we cannot afford to be expansive. We must be focused, and we must achieve returns on research investment. We face similar challenges to other countries, demonstrated by a recent EU report on science and policy-making.[5] The report highlighted the need to ensure that EU-funded research results inform policy-making in a meaningful way. EU policy-makers expressed a desire that stronger linkage should enhance the contribution of research to areas of major economic, social and scientific relevance for the EU.

For a decade or more the New Zealand government's principal funding agency, the Foundation for Research, Science and Technology (FRST), has had as a core principle that the public good research it funds must make a demonstrable contribution to outcomes of national value. Therefore research funding is targeted at projects that can show the pathway from research to such outcomes. This requires transparency around two areas in particular: the valuation of the outcomes and the pathway to uptake of research. Research users include government agencies and businesses, but also include non-governmental organisations, community groups, and other researchers.

FRST's assessment criteria for research proposals that range from €200,000 to upwards of €10 million are illustrated in Fig. 1.[6] Valuation of outcomes takes

---

[5]Directorate-General for Research, Socio-economic Sciences and Humanities (2008). Scientific evidence for policy making EUR22982 EN. Available at: http://ec.europa.eu/research/social-sciences/policy-publications_en.html.

[6]For an example of the funding portfolios and description of assessment criteria. Available at: http://www.frst.govt.nz/files/RfP%20Part%201%20Infrastructure%20Communities%20and%20Energy%20(ICE).pdf.

Benefit to New Zealand	Risk Management or success factors
1. Outcome benefits to New Zealand **Key question:** Assuming this project is successful, what is the potential contribution it will make to the achievement of target outcomes? **What should be included in your answer:** a) Assessment of the opportunity or need and potential impacts from the proposed research (for example, improved returns, avoided costs – economic, social, and environmental). b) The potential contribution to target outcomes, assuming successful adoption and implementation (including potential spill-over benefits). c) The research is beyond "business as usual" and is not likely to be undertaken by others in the near future (industry and private sector, public sector, community/voluntary).	**2. Implementation pathway** **Key question:** What is the likelihood the team will successfully have the research implemented? **What should be included in your answer:** a) Relevant team track record in delivering contributions to outcomes and working with research users (including, but not limited to, past Foundation contracted milestones). b) Barriers to adoption: • freedom to operation including, ethical issues, intellectual property and regulatory hurdles • takes account of existing or new competitors, substitutes and alternative ways of achieving outcomes. c) Pathway to research use: • plan (including mechanisms and milestones) to achieve uptake, capturing benefits for New Zealand • quality engagement with research users, and/or co-investment.

Benefit to New Zealand	Risk Management or success factors
3. Research, science and technology benefits to New Zealand **Key question:** Will the research be of high science quality and build or retain capabilities of potential future benefit for New Zealand? **What should be included in your answer:** a) Advanced knowledge in the research area. The magnitude of the extension of knowledge frontiers and generation of significant new knowledge, potentially contributing to the target outcome. b) Excellence of RS&T fit-for-purpose. Demonstration of soundness of research method (including research questions and/or hypotheses) and the rationale for the approach and methodology, and identification of the strengths/weaknesses of the research approach. c) The potential for this distinctive RS&T capability to contribute to target outcomes in future. Or d) How the maintenance and development of an existing RS&T capability makes, or may make, an important contribution to target outcomes, taking account of the difficulty of replicating it. (RS&T capability includes development of individual and team research abilities and associated support and infrastructure.)	**4. Ability to deliver research, science and technology results (outputs)** **Key question:** What is the likelihood the team will achieve their proposed research outputs? **What should be included in your answer:** a) RS&T track record of teams: • fit-for-purpose science track record • collaborations where necessary to bring together the best team to deliver the proposed research • track record of delivering on research (including, but not limited to, previous Foundation contracts). b) Project management and support: • team leadership, roles and coordination • organisational and infrastructural support • succession planning. c) Project plan: • clearly defined, well-sequenced critical path milestones and objectives • change events and identification of and mitigation of key project delivery risks.

Fig. 1 Assessment criteria for proposals used by the New Zealand government's research funding agency, the Foundation for Research Science and Technology in the Public Good Research fund. Target outcomes are set by the Foundation for portfolios of research in which funding is awarded

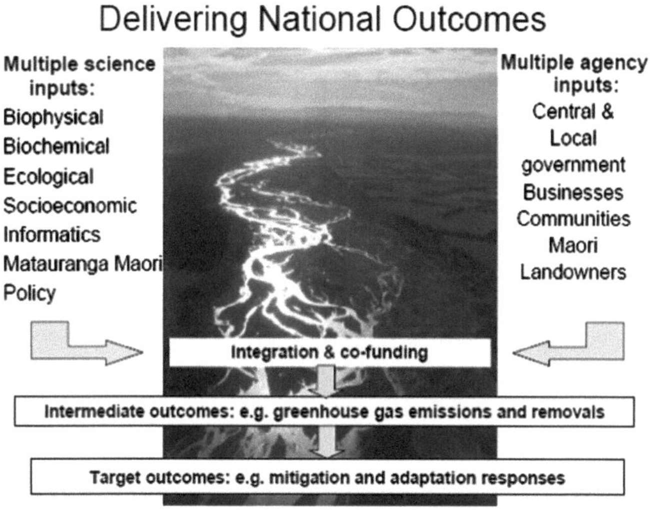

Fig. 2 The 'braided river' metaphor for integrating scientific disciplines and research user organisations in a fluid project structure to deliver intermediate and target outcomes of national benefit

a pragmatic approach by pointing to established national strategies or those of sector groups (e.g. dairy sector) that have been endorsed by government. Valuation may also include estimates of the economic value of outcomes (e.g. greenhouse gas research reducing economic liability under the Kyoto Protocol). Demonstrating the achievement of value may be problematic when, as is common, benefits are obtained after the project funding has finished. But it is possible to show that research has influenced policy and action in line with the intentions of the research proposal.

The pathway to uptake of research starts at the conception of the research programme. Evidence is expected by the funding agency of engagement between researchers and research users through the gestation of the project proposal, and this engagement may be audited by the agency when assessing the proposal. Engagement during the research project is the subject of contractual agreements with the funding agency. Researchers are bound by contract to deliver workshops, training programmes, publications, secondments, etc., to achieve research uptake. Research users may be bound by the same contract or a derivative, to fulfil their role in the pathway to uptake. Research programmes therefore bring together not only different disciplines, but also different research users, who may co-fund components in parallel, to achieve intermediate and target outcomes of benefit to New Zealand (Fig. 2).

Research is a partnership that is best fulfilled when the team includes both researchers and research users, supported by people with a range of additional skills. Figure 3 shows the stages in a conceptual research cycle together with the skills needed to enhance the value of the research at each stage. Beyond the essential skills in science and in research management, skills are needed in *translation* (both ways between the languages of science and users, e.g. policy-makers, and

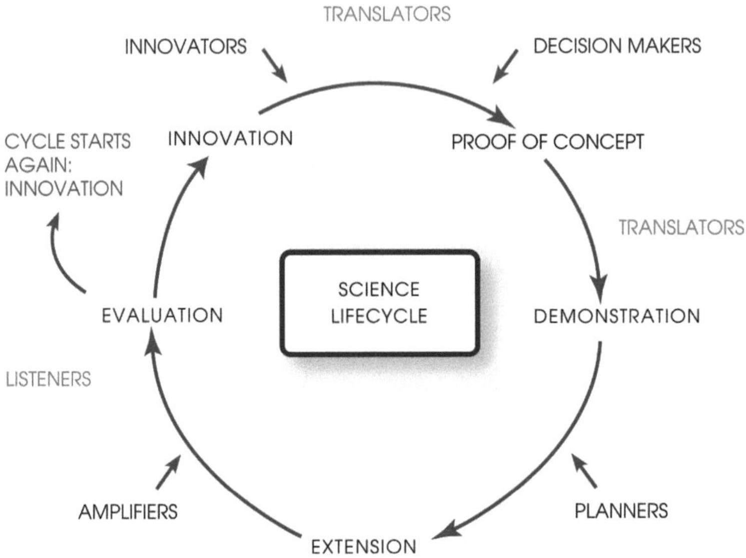

Fig. 3 Conceptual cycle of research including skills of translation to and from research users

funders); in *decision-making* (when to increase or decrease research funding, or take a different approach); in *planning* for the longer term implementation of research findings and tools beyond the funding lifetime; in *extension* or *amplification* of research from case studies to the mainstream; and in *listening, evaluation* and *collaborative learning* about the impacts of the research in its social context, from which may spring the new ideas that start the cycle again.

4 What Are Priorities for Investment in Sustainable Development Research?

The breadth of the subject defies simple analysis. Priorities for New Zealand like other countries extend across a wide spectrum from those deeply socio-economic to those deeply cultural and environmental, with all dimensions represented in most priorities. Figure 4 depicts a view of how water issues overlap nested economic, social, cultural and environmental dimensions. For example, issues of water consumption and allocation touch on all four dimensions; and Mauri (the Māori term signified by health and life force) connects the environmental and cultural dimensions. This approach helps to break down the silos in our thinking. Economic development, Māori affairs, climate change, and water are prominent in the present New Zealand government's agenda, and all relate to the complex challenge of

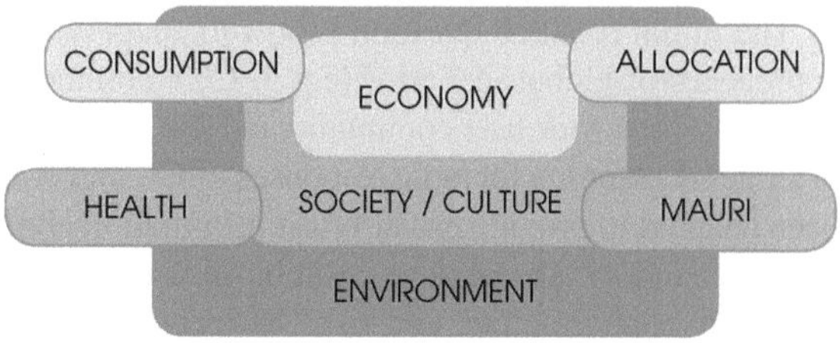

Fig. 4 Concentric domains of environment, society and economy overlaid with water issues. (*Note*: Mauri is a Māori word meaning life-force)

achieving development that sustains and grows the social, environmental and cultural resources on which it depends.

In a time of great uncertainty about the future *governance for sustainable development* is a particularly relevant theme. Governance, rather like sustainability, is a term with multiple meanings. In the context of this paper the hallmarks of governance are those of effective Boards of Directors: attention to vision and longer-term strategy, risk and opportunity, relationships with stakeholders, goal-setting, and overseeing prudence in management. Governance here relates to both business and government.

Let us explore a research agenda on governance for sustainable development, with four examples providing a New Zealand perspective:

- Futuring for agile organisations
- Resilience and adaptive capacity in communities
- Post-regulatory governance for constrained natural resources
- Governance models from indigenous communities

4.1 *Futuring for Agile Organisations*

Change is speeding up, increasing the pressures on central and local government to provide 'agile' responses to increasingly 'wicked' problems.[7] These are multidimensional, with messy solutions, in which uncertainty and risks are typically high, and often there is no "right" answer.[8] Yet agile responses are required when investing strategically in infrastructure, business, and human capital against a global backdrop of significant and uncertain political, economic, social and environmental change.

[7]Demos W. J., & State Services Authority (2008). Towards Agile Government, State Services Authority, Victoria, Australia. Available at: http://ssa.vic.gov.au.

[8]Rayner, S. (2006). Wicked Problems, Clumsy Solutions: Diagnoses and Prescriptions for Environmental Ills. Available at: www.martininstitute.ox.ac.uk/NR/rdonlyres/C3EDD045-9E3B-4053-9229-9CF76660AAC6/645/JackBealeLectureWickedproblems.pdf.

The initial research question is how to adapt and combine three common futuring approaches of global scenarios (e.g. IPCC), community visioning, and New Zealand scenarios, for a wide user community in New Zealand and so improve the effectiveness of strategic planning for agile responses. IPCC and other global climate scenarios have been adapted to provide broad-brush information about likely climate changes within New Zealand's major regions.[9] But businesses and government still lack the capacity to identify opportunities and risks to specific organisations or communities. Local government legislation has produced Long-term Council Community Plans,[10] but tools are only now in development to give Councils and communities the capacity to model the implications of alternative policies for integrated environmental, social and economic outcomes. One example in New Zealand is the Creating Futures programme[11] which builds on technology sourced in Europe.

An example of national futuring is Four Scenarios for New Zealand. The four scenarios (named *New Frontiers*, *Fruits for the Few*, *Independent Aotearoa*, and *Living on Number8 Wire*) occupy a matrix with axes of identity (individuality – cohesion), and resources (plentiful – highly constrained) (Fig. 5). They give a rich sense of how life could differ in the future: at work, at home, in politics and in business. With whom will we trade? What sports will we play? How will we educate people? And what will all this mean in terms of sustainable development? Since these scenarios and a futures "game" derived from them were developed, 34 organisations in central and local government and the private sector have been enabled to take the long view and explore futures thinking in parallel with strategy exercises.

In spite of those initiatives, contemporary futuring risks being a separate exercise, not mainstreamed in strategic planning or community debate. In a series of workshops and interviews on research directions in 2007–2008 a consistent message from research users was the need to address New Zealand's lack of capacity in translating futures into strategy. We identified three opportunities: first, to improve alignment between future scenarios and government policies such as regional development form, transport, infrastructure provision, and natural resource governance; second, to align global economic and social trends with policies for labour and human development; and third, the use of futures by businesses in re-modelling to address environmental and social opportunities, especially as organisations orient themselves into a new world order post-recession.

Two initiatives have the potential to support the capacity for such alignments between futures and strategy. First is to create a shared understanding and resource base of future scenarios relevant to New Zealand. This has the potential to improve the quality of strategic planning, reduce the inevitable duplication of effort between agencies needing such knowledge, and support those with inadequate resources

[9]NIWA (2009). Regional modelling of New Zealand climate. Available at: http://www.niwa.co.nz/our-science/climate/research-projects/all/regional-modelling-ofnew-zealand-climate.

[10]Long-term Council Community Plans are required under the Local Government Act 2002. An example is that of Environment Waikato. Available at: http://www.ew.govt.nz/policy-and-plans/Long-Term-Council-Community-Plan-Annual-Plan-and-Annual-Report/.

[11]Creating Futures http://creatingfutures.org.nz/.

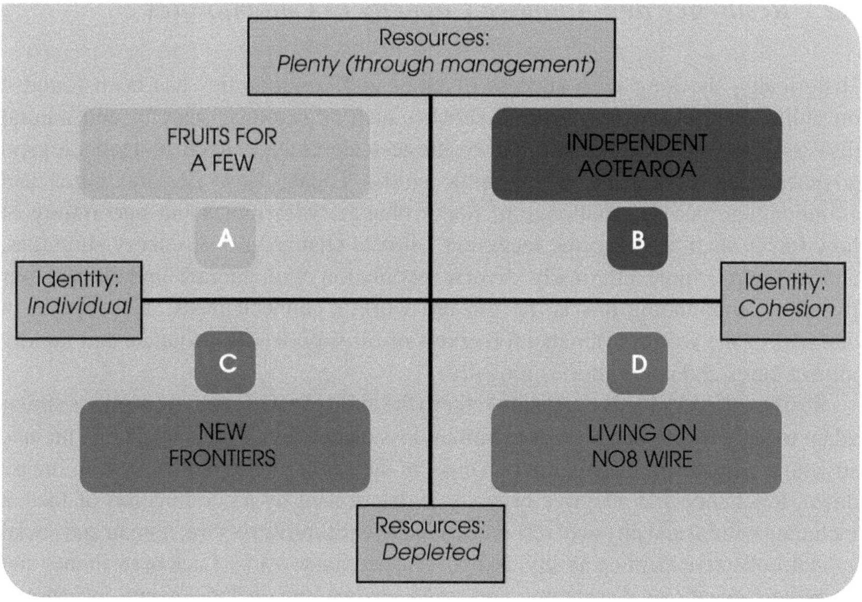

Fig. 5 Four scenarios for New Zealand[12]

or capacity for doing effective futuring. A deliverable in the pathway to uptake is to put leading international resources on future pressures and opportunities "on every desk" in government (and other sectors), including new methods of engaging citizens in ongoing debate about future scenarios using Web 2.0 and 3.0 technologies, as has been started by the European Commission.[13]

The second initiative is to create a Virtual Futures Laboratory in which government, business and other organisations can explore strategic options using relevant science and research tools in a "safe" space. The aim with this initiative is to overcome the barriers to accessing and using research and science that have been encountered by policy-makers and businesses. Barriers include science being narrow and deep, and in silos rather than broad and integrated in a way that reflects the realities of policy-making and business strategy. An effective Virtual Futures Laboratory would make tools and expertise available in a joined-up way, allowing "experiments" that explore alternative futures and innovations, without putting communities, economies and environments at risk.

---

[12]*100% Pure Conjecture', a* participative game to stimulate interest in future directions for New Zealand and to aid strategic-thinking about sustainability. Available at: http://www.landcareresearch.co.nz/services/sustainablesoc/futures/about.asp.

[13]http://www.europa.eu/debateeurope/ is used by the European Commission to actively listen and engage in dialogue with its citizens.

4.2 Resilience and Adaptive Capacity in Communities

Historically, the long term success of cities and communities has been founded on ability to prevent or withstand shocks, such as resource scarcity and natural disasters, and adapt and capitalize on large-scale change, such as technological advances and significant demographic shifts. Today, New Zealand cities and communities face the challenge of major change, with increasing uncertainty of how forces such as economic recession, climate change, global energy shortages, and an aging, more ethnically diverse population will interact and impact our lives.[14] Compounding this is the modern world's connectedness; a disruption in one part of the world, to financial markets or oil supplies for instance, can rapidly impact cities and communities globally.

Resilience and adaptive capacity refer to the ability to withstand disruptions and/or adapt to large scale change with minimum loss of function. The concept can include structural adjustment or structural change, in the event of substantive system breakdown. Resilience and adaptive capacity is determined by a combination of factors including natural and physical resources, character of infrastructure, human and social capital, collective learning ability, and governance frameworks. Lack of resilience and adaptive capacity to disruptions and rapid change can include major job losses; deterioration of natural resources; capital losses from obsolescence in buildings, roads, and plant; the breakdown of critical infrastructure systems; social dislocation; and losses in personal and cultural identity. The aim of research is to show how such costs can be replaced with net benefits from, for example, designing adaptable infrastructure and flexible building systems, positioning communities to gain from emerging economic sectors, and strengthening community and business competitiveness with a culture of preparedness and environmental leadership.

In order to build resilience and adaptive capacity, we need to understand what factors and processes make some settlements vulnerable to disruptions and rapid change while others can adapt.[15,16] The desired national outcome is to enable local and central government to build this capacity, moving beyond the current focus on crisis events and disaster management. A framework, indicators and community planning toolkit are needed to enable New Zealand city managers and central government agencies to work with communities and gain their mandate in implementing proactive management responses to uncertain futures.

Local councils will make use of an adaptive communities planning toolkit in preparation of the 2015 Long-term Council Community Plans[13] and in structure and infrastructure plans, which address impacts on future generations. Spill-over benefits are anticipated in communities adopting new economic activities and

---

[14]Auckland Sustainability Framework (2007). Overarching strategic framework for local and central government decision-making in the Auckland Region. Available at: www.sustainingauckland.org.nz.

[15]The urban resilience prospectus CSIRO, Australia, Arizona State University, USA, Stockholm University, Sweden www.resalliance.org/1610.php.

[16]CitiesPlus: www.citiesplus.ca/.

creating new jobs – with greater diversity being an adaptive response to uncertain futures.

4.3 Post-regulatory Governance of Constrained Natural Resources

Sustainable use of natural resources is the foundation for primary industries that play a major role in New Zealand's national and regional economies. Dairy and meat products alone account for 33% (NZ$10 billion) of export income. Hydroelectricity provides over 60% of New Zealand's electricity, while other renewable energy resources are increasingly important. Equally, New Zealand's unique and spectacular environment is a primary draw card for international tourism, which accounts for 18.5% (NZ$7.4 billion) of exports. The success of these and other industries depends in large part on their access to, and use of, high quality natural resources that are becoming increasingly scarce.

Apart from the economic value of natural resources, integrity of natural systems is of increasing concern to New Zealanders. Economic and other resource uses and values are increasingly coming into conflict, creating difficult problems of natural resource governance. Conflicts over water allocation are increasing, as are problems of water pollution. Development of alternative energy resources is often contentious, as are many coastal developments.

In these and many other cases, there are important and contested issues around what is physically, legally, economically, and socially feasible, and then what is desirable, in the management of common resources. Furthermore under resource management and local government legislation, local authorities have a responsibility to recognise the incorporation of Māori perspectives in planning and decision-making, but often struggle with how to implement this effectively.

Successful natural resource governance can only be achieved through integration of social, environmental, economic and cultural dimensions. Present decision-making typically moves from an imperfect regulatory environment to a combative legal environment in the Courts. Attention is becoming focused on the opportunity for post-regulatory approaches that incorporate stakeholder collaboration, consensus building, and more integrative, inter-disciplinary research.[17]

A research agenda we are following is the development of an integrative framework for analysis of natural resource governance problems in terms of efficiency, effectiveness, equity, legitimacy and scale.[18] The research has taken

[17]Gunningham, N. (2008). Innovative governance and regulatory design: managing water resources. Landcare Research Contract Report: LC0708/137. Available at: http://www.landcareresearch.co.nz/publications/researchpubs/water_gunningham_LC0708137.pdf.

[18]Adger, W. N., Brown, K., Fairbass, J., Jordan, A., Paavola, J., Rosendi, S., & Seyfang, G. (2003). Governance for sustainability: towards a "thick" analysis of environmental decisionmaking. *Environment and Planning A*, *35*, 1095–1110.

an initial focus on water, but the framework and methods could be applicable to natural resource governance in many sectors and regions of New Zealand. The research draws on a wide range of scientific disciplines, using both quantitative and qualitative methods.

Quantitative models are being developed at both regional and local scales to create better understanding of the role of water in economic production. An 'integrated computable general equilibrium' model has been developed, capable of simulating the broad effects of alternative policies and alternative scenarios for economic development at the regional scale.[19] An 'agent based model' will also be developed to explore specific issues in more detail at the scale of multiple catchments.

Qualitative approaches are being used to develop a better understanding of decision-making processes around sustainable allocation and use of water resources. We are producing an institutional landscape map by examining the legal and institutional frameworks; exploring informal, or 'silent', accounts of experiences of inter-agency decision-making processes, including aspects of authority and institutional barriers to creating new mechanisms of regional planning; examining media representation of water issues; and analysing relevant policies from within and beyond New Zealand's shores.

Collaborative learning techniques build capability in stakeholder engagement and constructive use of scientific knowledge. Where these techniques focus on Māori issues and perspectives, Māori researchers establish and articulate Māori perspectives and knowledge on resource issues and identify appropriate governance models. This often involves finding out how stakeholders understand and interpret the 'Māori voice' with respect to natural resource governance and recommending equitable New Zealand solutions.

4.4 Governance Models from Indigenous Cultures

The first humans arrived in Aotearoa-New Zealand from Polynesia about 800 years ago, populated the country, and evolved a distinct Māori culture inextricably linked with the natural and spiritual environment. Europeans first settled in New Zealand in the early 1800s, and the Treaty of Waitangi was signed with Māori chiefs in 1840 to provide Māori rights over their lands, resources, and taonga. However, under European colonisation, an intense period of Māori land alienation and confiscation of strategic resources followed until about 1940 when Māori land represented only 6% of Aotearoa New Zealand. A new era commenced in 1975 in which the Crown (New Zealand Government) recognised the resource alienation as a significant historical grievance, and entered a phase of dialogue, dispute resolution, and settlement.

[19] Lennox, J. A., & Diukanova, O. (2008). Modelling regional general equilibrium effects and irrigation in Canterbury. International Conference on Policy Modelling (Ecomod 2008), Berlin, 2–4 July 2008. Available at: http://www.landcareresearch.co.nz/research/programme_pubs.asp?Proj_Collab_ID=94.

The resulting compensation to Māori tribes for land and economic losses has provided many with the opportunity to once again govern significant assets and resources (e.g., land, fisheries, property) and to build an economic, social, and cultural base on which to develop a sustainable future for their people. Indigenous Māori make up about 15% of New Zealand's population of 4 million, with about 80% of all Māori now living in urban areas. The 2010 Māori commercial asset base was worth NZ$39.6 billion (a real growth of 4.3% per annum from 2006) and the value added to the economy by Māori enterprises totalled NZ$10.3bn. 52% of Māori commercial assets are concentrated in primary industry such as farming, forestry, fisheries, and agriculture, while 40% is in the tertiary sector, representing growing numbers of Māori who are self-employed and entrepreneurs.[20] A significant question for many Māori organisations and businesses has been how to balance aspirations for cultural enrichment (e.g., retaining strong elements of traditional culture such as values, language and knowledge) with more modern elements of advancement, growth, commerce and economic development (Harmsworth 2006). Our research with a number of Māori businesses[21,22] has shown that effective corporate governance is a necessary precursor to integrating cultural heritage and values into an organisation. It is also essential to have a robust organisational planning and reporting framework, in which to articulate goals and outcomes, and implement, measure and report performance. Our research seeks to support that development of governance.

Durie[23,24] posed the broad question "how is a Māori business distinguished from any other business?" He identified the following six key outcomes that could be used to evaluate a Māori business's contribution to Māori development and advancement: Tūhono (aligns a Māori business to Māori aspirations through comprehensive consultation); Pūrotu (transparency and responsibility to the wider community); Whakaritenga (balanced motives, not just profit-making); Paiheretia (integrated goals, using effective management); Puāwaitanga (best outcomes within wider social, cultural, environmental and economic, perspectives and goals); and Kotahitanga (unity and alliance that encourages cooperation).

These elements distinguish emergent Māori business. They also define a governance framework that has relevance in a world seeking a new social contract between business and society. They look to the long-term sustainable future:

[20]Te Puni Kōkiri, BERL (2011). The asset base, income, expenditure and GDP of the 2010 Māori economy. Wellington: Te Puni Kōkiri.

[21]Harmsworth, G. R. (2006). Governance systems and means of scoring and reporting performance for Māori businesses. Landcare Research paper for Mana Taiao, Foundation for Research Science & Technology (2003–2007). Available at: http://www.landcareresearch.co.nz/research/sustainablesoc/social/indigenous_index.asp.

[22]Harmsworth, G. R., & Tahi, M. (2008). Indigenous Branding: Examples from Aotearoa – New Zealand. 22–25 July 2008. FIBEA – Fostering Indigenous Business & Entrepreneurship in the Americas Conference, Manaus, Brazil. Accepted April 2008 for Conference Proceedings.

[23]Durie, M. (2002). The business ethic and Māori development. Paper presented at Maunga Ta Maunga Ora Economic Summit March 2002, Hawera.

[24]Durie, M. (2003). *Ngā kahui pou: launching Māori futures*. Wellington: Huia Publishers.

"Mō tātou, ā, mō kā uri ā muri ake nei" (for us and our children after us),[25] and they express the spirit of sustainable development: "Manaaki Whenua, Manaaki Tangata, Haere whakamua" (Care for the land, Care for the people, Go forward – We are the guardian of our assets and community).[26]

5 Conclusion

The Land of the Long White Cloud, Aotearoa New Zealand, may not yet have the answers to sustainable development, despite our 100% pure, clean-green image. But we have a pragmatic approach to developing research agendas in *partnership* with research users; our country is a *national laboratory* for solutions of relevance to other countries; a *long view* and *futures* have the potential to inform our policy and strategy across sectors; we can learn from the economic crisis to create *agility, resilience and adaptive capacity* in our organisations and communities; and *Māori values and practices* are helping fashion distinctive approaches towards equitable societal goals for sustainable development in this generation and beyond.

6 Epilogue (November 2010)

In the General Election of November 2008, leadership of the New Zealand government changed from the centre-left to centre-right. Sustainability had been a major theme of the Labour-led coalition government.[27] The new National-led coalition government came into power in the aftermath of the global economic crisis and set an Economic Growth Agenda (EGA) for its term in office.[28] Government departments and science expenditure were to be aligned with this agenda. The EGA encompassed six main policy drivers: a growth-enhancing tax system, better public services, support for science, innovation and trade, better regulation, including regulations around natural resources, investment in infrastructure, and improved education and skills. Science and technology were seen to be part of the solution to economic challenges. The "sustainability" word was largely removed from the government's language.

In 2009 the research programme, *Building Capacity for Sustainable Development: the Enabling Research* came to the end of its 6-year government funding. Proposals to continue and further develop research on futures and resilience

[25]Ngai Tahu (2009). Available at: http://www.ngaitahu.iwi.nz/About-Ngai-Tahu/.

[26]Wakatu (2009). Available at: http://www.wakatu.org/main/Vision_and_Values/.

[27]In 2007 Prime Minister Helen Clark declared, "I believe New Zealand can aim to be the first nation to be truly sustainable – across the four pillars of the economy, society, the environment, and nationhood".

[28]Economic Growth Agenda. Available at: http://www.med.govt.nz/templates/Page_44545.aspx.

(Sects. 4.1 and 4.2 above) were submitted in the national contestable funding process and were unsuccessful.

Also in 2009 the New Zealand government established a taskforce to recommend changes in the focus and funding of the eight government-owned Crown Research Institutes (CRIs).[29] In 2008/09 the total revenue of the eight CRIs was NZ $675 million, they accounted for a quarter of New Zealand's total research expenditure and employed a combined staff of 4,400.

The taskforce recommended that greater clarity be achieved in the core purpose and scope of the CRIs to reduce overlap and perceived wasteful competition in New Zealand's highly contestable research funding process. It also recommended that government should directly fund CRIs for a significant proportion of their core purpose work, which would remove contestability for that funding in return for clear accountability for engagement with users and contributions made to achieving national outcomes.

A Statement of Core Purpose was developed for each of the eight CRIs, which cover industrial, primary sector and environmental research. A CRI's core purpose is defined by the national outcomes to which its science should contribute; and its scope is defined as areas of science on which it focuses.

Neither *sustainable development* nor the emerging term *green growth*[30] is used in the Statements of Core Purpose; but terms that are used include prosperity, environmental sustainability, development within environmental limits, sustainable management, the protection and enhancement of terrestrial resources, and integrated social and biophysical research.

One national outcome in the SCPs is "*industries developing within environmental limits and responding to market and community requirements.*" This outcome reflects a growing interest by the business sector, exemplified by Business New Zealand,[31] in the impacts of constrained resources (both environmental and social) and increasing pressures for performance and transparency from local communities and overseas markets – not only in Europe.[32]

The role of science contributing to this national outcome in New Zealand has developed over the last decade with the CRI, Landcare Research, and others creating knowledge and tools to assist businesses in responding to climate change, greenhouse gas emissions, resource management issues, and sustainability reporting.[33]

---

[29]CRI Taskforce report. Available at: http://www.morst.govt.nz/current-work/CRI-Taskforce/Final-Report/.

[30]Green Growth strategy, see www.oecd.org/greengrowth.

[31]Business NZ. "Why take the lead on sustainability?". Available at: http://www.businessnz.org.nz/content/BusSustLead.pdf.

[32]See Walmart's Supplier Sustainability Questionnaire. Available at: http://walmartstores.com/download/4055.pdf.

[33]See the e-book *Hatched: the capacity for sustainable development*. Available at: http://hatched.net.nz.

There is a growing recognition by businesses of the importance of sustaining and enhancing the environmental assets, which underpin a national economy that is largely based on the primary sector and tourism.[34] But there is also an emerging recognition of the opportunities for business to innovate and compete in markets where the natural environment is a key factor – either because markets demand improved environmental performance, or because they seek products and services that address environmental issues.[35] Our challenge and opportunity in New Zealand are to align our science with these trends.

While business culture is changing to recognise the environment as an asset, Māori culture is already founded on that recognition and goes further to recognise the inter-connectedness of mankind and all aspects of the natural world (see Sect. 4). In Aotearoa New Zealand this recognition is embodied in the term, Mātauranga Māori, which integrates a diversity of types of traditional knowledge (tacit, scientific, religious, etc.) on a range of human activities, spiritual domains, and ethics. Mātauranga Māori is an important aspect of indigeneity, which plays an integrating role across the world's cultures.

Our challenge and opportunity in New Zealand are to bring "western" science and Mātauranga Māori into one flow, like rejoining rivers separated in the Enlightenment of the eighteenth century. In the sense that science is increasingly integrated (across disciplines) and integrating (across stakeholders) it is convergent with Mātauranga Māori, which is founded on inter-connectedness. In sustainability science the concept of ecosystem services reflects linkages between the environment, society and economy. But the word "services" does not reflect the reciprocal relationship understood by Māori.

One model for advancing beyond the utilitarian notion of ecosystem services has been finding linkages between scientific and cultural indicators of waterway health in a New Zealand catchment.[36] Scientific methods measure precise changes to river health over time but can be relatively costly and require a high degree of professional expertise and experience. The cultural approach provides qualitative, cost effective detection and measurement of long-term changes. It is based on acquiring in-depth knowledge of a local environment through mātauranga Maori, community and historical knowledge. The combination of approaches provides an enriched understanding of the environment through differing worldviews about the health of freshwater systems.[37]

---

[34]Gordon, R.F.S. (2010). *The environment as an economic asset*, presentation to the Royal Society/Science NZ Parliamentary Speaker's Science Forum, May 2010, on http://www.landcareresearch.co.nz/sustainability/sustainabilty_details.asp?Sustainability_ID=121.

[35]Green Growth in New Zealand. See for example the *100 per cent Plan* http://www.100percentplan.com/.

[36]Integrated catchment management project in the Motueka Valley, Tasman, New Zealand. See http://icm.landcareresearch.co.nz/knowledgebase/publications/public/Cultural_indicators_report2.pdf.

[37]Watershed Talk: a project about processes for building community resilience. Available at: http://icm.landcareresearch.co.nz/knowledgebase/publications/public/Watershed_Talk_Summary_2009.pdf.

In addressing the world's "wicked problems" of sustainable development, science and Mātauranga Māori can provide answers to questions like, "*what do we know?*" and "*what questions should we ask?*" But this leaves open the question of what role should be played by the science community, in whatever ways scientists associate – as institutes, universities, wānanga (Māori learning institutions), global networks, etc.?

In 2010 we observe a growing recognition for the role played by the science community in global diplomacy by occupying "common ground" between countries and providing an entry point for new relationships. For example, as the global climate change negotiations at Copenhagen struggled to make commitments to reduce emissions, a scientific collaboration led by New Zealand, the Global Research Alliance on Agricultural Greenhouse Gas Emissions, offered a constructive way forward.[38] Thirty countries have joined the Alliance and will find ways to align research programmes, share resources and speed up progress with the intransigent problems of methane and nitrous oxide emissions and understanding soil carbon dynamics.

The Alliance leadership confronted the reality that slow progress was being made with the challenge of increasing food production while reducing greenhouse gas emissions from the production process. A vision was promoted of enhanced interdependence in science efforts and optimism about building capability and achieving goals through the exchange of ideas and researchers. The Alliance has the potential to raise energy levels and freedom to mobilise greater resources in member countries.

Such leadership can be a model for the science community in all areas of sustainable development, confronting the perceived realities in society of "climate change fatigue" and despair that global problems of food security, resource depletion, and social deprivation are too hard. At a time when leadership can be thought of as increasingly distributed through the impacts of social networking technology, the global science community can start new conversations between world views, enhance capabilities in society, and inspire self-motivation across those sectors that have a role to play.[39]

Acknowledgements and Disclaimer This paper reflects a small selection of research projects conducted or conceived by Landcare Research, a Crown Research Institute in New Zealand.[40] The author especially recognises the contributions of research leaders: Bob Frame, Garth Harmsworth, Michael Krausse, James Lennox, Claire Mortimer, and Daniel Rutledge; and both the Ministry and the Foundation for Research, Science and Technology who funded some of the research described here. Views expressed in the November 2010 Epilogue are those of the author. This paper is not intended to represent views of the New Zealand government.

---

[38]See http://www.globalresearchalliance.org/.

[39]Gordon, R.F.S. (2009). Unending in *Hatched: the capacity for sustainable development*. http://www.landcareresearch.co.nz/services/sustainablesoc/hatched/overview.asp#s5.

[40]See www.landcareresearch.co.nz. Landcare Research's purpose is to drive innovation in New Zealand's management of terrestrial biodiversity and land resources in order to both protect and enhance the terrestrial environment and grow New Zealand's prosperity.

Integrated Water Resources Management: STRIVER Efforts to Assess the Current Status and Future Possibilities in Four River Basins

Per Stålnacke, Geoffrey D. Gooch, Udaya Sekhar Nagothu, Ingrid Nesheim, Line J. Barkved, Bruna Grizzetti, Alistair Rieu-Clarke, Johannes Deelstra, Haakon Thaulow, Dag Berge, Antonio Lo Porto, Dang Kim Nhung, S. Manasi, and Santiago Beguería Portugués

P. Stålnacke (✉) • U.S. Nagothu • J. Deelstra
Bioforsk – Norwegian Institute for Agricultural and Environmental Research Norway
e-mail: per.staalnacke@bioforsk.no; usn@bioforsk.no; johannes.deelstra@bioforsk.no

G.D. Gooch
Linköpings University, Linköping, Sweden
and
IHP-HELP Centre for Water Law, Policy and Science (under the auspices of UNESCO), University of Dundee, Dundee, Scotland, UK
e-mail: geoffrey.gooch@liu.se

I. Nesheim
The Centre for Development and the Environment, University of Oslo, Norway
e-mail: ingrid.nesheim@sum.uio.no

L.J. Barkved • H. Thaulow • D. Berge
NIVA – Norwegian Institute for Water Research, Oslo, Norway
e-mail: hth@niva.no; lba@niva.no; dag@niva.no

B. Grizzetti
Joint Research Centre of the European Commission, Ispra, Italy
e-mail: bruna.grizzetti@jrc.it

A. Rieu-Clarke
IHP-HELP Centre for Water Law, Policy and Science (under the auspices of UNESCO), University of Dundee, Dundee, Scotland, UK
e-mail: a.rieuclarke@dundee.ac.uk

A.L. Porto
Water Research Institute – National Research Council, Bari, Italy
e-mail: antonio.loporto@ba.irsa.cnr.it

D.K. Nhung
Institute of Geography, Hanoi, Vietnam
e-mail: nhungtuti@yahoo.com

S. Manasi
Institute for Social and Economic Change, Bangalore, India
e-mail: manasi@isec.ac.in

S.B. Portugués
Consejo Superior de Investigaciones Científicas, Zaragoza, Spain
e-mail: sbegueria@eead.csic.es

C.C. Jaeger et al. (eds.), *European Research on Sustainable Development*,
DOI 10.1007/978-3-642-19202-9_11,
Published by Springer-Verlag Berlin Heidelberg 2011

1 Introduction

The contemporary concept of Integrated Water Resources Management (IWRM) was primarily conceived for the purpose of promoting sustainable water management. There are many elements included in modern IWRM perceptions, e.g., natural resource utilization planning combined with at strategy to balance between social, economic and environmental objectives based on an overall sustainability concept. However, the concept behind IWRM is not new. The historical development of the concept of Integrated Water Resources Management (IWRM) can be found in Rahaman and Varis (2005). The Technical Advisory Committee of Global Water Partnership (GWP-TEC 2000) has adopted the following definition of IWRM that has so far, received the most citations:

> IWRM is a process, which promotes the coordinated development and management of water, land and related resources, in order to maximise the resultant economic and social welfare in an equitable manner without compromising the sustainability of vital ecosystems.

The European Union Water Initiative (EUWI) launched at the Johannesburg Summit in 2002 had the overall objective to promote the implementation of IWRM based on a river basin approach and to support – as advocated in the Johannesburg Plan of Implementation – the adoption of water resource management strategies and plans by 2005. As a follow-up to this, the European Commission (EC) via Directorate-General Research launched a "twinning" mechanism in Framework Programme 6 with the overall goal that it should be a science-based contribution to the EUWI. The aim was to "twin" European river basins with basins outside Europe to facilitate a platform for the sharing of experiences around IWRM. So far seven twinning projects have been introduced, one of which is STRIVER.

This chapter first provides an overview of the STRIVER project, followed by a selection of overall findings in an IWRM-setting, before providing a final discussion on the implication of the findings in the context of research and sustainable development. To large extent this chapter is a summary of two recently launched text-books on sustainable water management derived from the STRIVER project (Gooch and Stålnacke 2010a; Gooch et al. 2010).

2 Methodology and Study Area

2.1 Methods

Given that water and river basin management demands a combination of information produced by different scientific disciplines, a particular goal of the project was to develop methods to integrate results from three different scientific standpoints; namely (1) a natural science perspective, involving studies of ecological flows, nutrient and sediment loads and its impact on ecology, (2) an information and knowledge perspective, involving e.g., collation of data and mechanisms to disseminate scientific information, and (3) a policy and social science perspective, involving studies of policy and legal instruments, water pricing and economic valuation, and stakeholder participation. It was assumed that a viable approach for IWRM involves the creation of a framework that combines inputs from the various scientific disciplines, policy and management communities as well as local water use communities.

More specifically, based on the development of a multidisciplinary knowledge base in all the case basins, and an IWRM conceptual framework, the project carried out an IWRM assessment in four case basins (see Sect. 2.2). Then specific problems or focus areas were covered; namely (1) transboundary water governance, (2) pollution modelling, (3) environmental flows, and (4) land and water use interaction. The target beneficiaries of the project results were – besides the research community-, water managers, basin authorities, policy makers, and water users in the case basins. Stakeholder-interaction, communication and dissemination aspects have been particularly strong in STRIVER (see Sect. 3) thus increasing the likelihood of producing tailor made recommendations. How this was done is further shown in Sect. 3. The partners have invested substantial efforts in order to understand the various "paradigms" and view-points raised by the local communities, managers and policy makers.

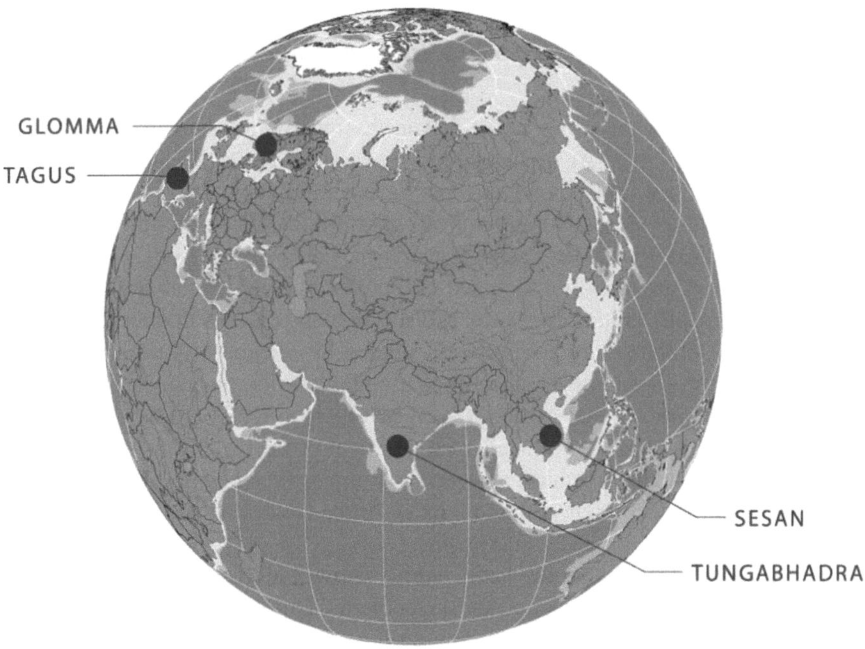

Fig. 1 The selected four river basins in STRIVER: Glomma (Norway), Tagus (Spain and Portugal), Tungabhadra (India) and Sesan (Cambodia and Vietnam) (*Source*: Gooch and Stålnacke (2010b))

2.2 Study Area

The four project river basins in which information, knowledge and competence were twinned, were the: Glomma (Norway), Tagus (Spain and Portugal), Tungabhadra (India) and Sesan (Vietnam and Cambodia) (Fig. 1). The basins where selected according to the following criteria: (1) transboundary, (2) pollution issues (3) hydropower development evident, and (4) varying degrees of IWRM implementation, including public participation and institutional frameworks. A more detailed description of the basins is given in the text-book by Gooch and co-workers (2010).

3 Results

In the following sub-sections, we first elaborate on the science-policy-stakeholder interface experiences gained in the project. This is followed by the IWRM status assessment in the four basins and a summary of various IWRM-topics addressed in STRIVER. The more detailed results can be found in two recently published text-books (Gooch and Stålnacke 2010a; Gooch et al. 2010).

3.1 Science-Stakeholder-Policy Interface and Scenarios

A major problem in the move to IWRM is the incorporation of different forms of knowledge into the part of the policy process known as the science-policy-stakeholder interface. This knowledge can be scientific, local, or a combination of both. Despite the considerable amount of effort put into research into the factors influencing IWRM, the results of this research, as well as inputs from scientific and local knowledge, are often not included into policy making (Gooch and Stålnacke 2010b). The reasons for this are many, and still in many cases uncertain. The initial formulation of problems, often a process steered by scientists, is usually heavily influenced by the need to develop or refine problem solving procedures that fit into scientific disciplines, or that can be judged by others in the scientific community. Policy makers and managers, faced with practical problems and the demands of their electorates, look for quick answers to immediate problems. Both of these approaches are perfectly relevant and logical for the concerned parties, but the gap between them often seems insurmountable.

The demands on stakeholder and public participation, which are a central aspect of IWRM, have attenuated these problems, and the question now is how to include inputs from these different groups in policy processes? How can stakeholders and the public contribute to policy making in such scientifically complicated fields such as water management and IWRM? Are stakeholders and the public able to understand the complexity of these issues? The results of the work undertaken in the STRIVER project show that the answer to this question is yes, under certain conditions they can. There are today a number of tools that enable stakeholders to participate in environmental management (Gooch and Huitema 2008) although often they facilitate discussions and comments on existing proposals more than direct input into scientific approaches. While these may enable non-experts to understand complex problems, they do not provide a means of providing stakeholders and the public with ways to provide direct input in planning for the future. Planning for the future is also a major challenge for the scientific community as a lack of information about future conditions makes modelling exercises dependent on expectation of future developments. There is always uncertainty. At the same time, the effects of successful or unsuccessful water management through IWRM can be quite simply a matter of life and death in many parts of the world. As such, those most affected by the management of this vital resource should morally be involved in discussions as to how best to manage it. One way of involving stakeholders and the public in the formulation of possible futures is through the use of scenarios. These are projections of possible futures (Alcamo 2001; Shell International 2003), not necessarily the most likely futures.

In STRIVER stakeholders and the public were included in the formulation of scenarios for sustainable water management on the Sesan (Vietnam-Cambodia) and Tagus (Spain/Portugal) rivers, and to a lesser degree in the other basins. Water management regimes were first examined and it was seen that the combination of law, policy, actors and institutions was vital to these regimes. Communication processes were also of central importance as it within these that information, knowledge, and mutual understanding of problems and their solutions, are formed. Law plays a vital role in conferring rights and obligations on actors in support of IWRM. However the formal adoption of appropriate laws is meaningless without securing their effective implementation and for this it is necessary with the support of stakeholders and the public. In STRIVER, qualitative scenarios were used as a means of involving stakeholders in the formulation of policies, as well as a way to improve social learning processes and the potential of policy implementation. As a first step, the most important influences on future water use were discussed in groups and stakeholder meetings. These drivers were then combined into four scenarios, which there then presented to the stakeholder groups for discussion, changes, and refinement. This process was repeated three times, so that the final scenarios were a combination of scientific and local knowledge. A major achievement of this interactive, participatory process was that groups with radically differing views were able to express their desired or feared futures for each other, something which then provided a base for discussion.

Obviously stakeholder participation and analysis is a crucial element for practical IWRM implementation. In the STRIVER project, stakeholder participation was given an important place as demonstrated through the series of stakeholder workshops that were conducted in all the four case basins each year. In all, 12 stakeholder workshops were conducted (three within each basin), in addition to a range of targeted discussions with key stakeholders. During the first year, the workshops provided a platform for reviewing the initial stakeholder analyses and identifying the key stakeholders; introducing stakeholders to the project objectives and mapping stakeholder expectations, interests and problems; and fostering synergies with ongoing activities within each of the basins. In the second year, stakeholder workshops were used to steer research objectives and activities, including developing policy scenarios. The final year workshops were essential as a tool for collaboratively reviewing project outputs with stakeholders and the STRIVER team, as well as identifying avenues for further exploitation of STRIVER results.

The workshops not only helped in the integration of various perspectives of stakeholders from different sectors, but also different user groups with varied and conflicting interests. Experience showed that there was a relatively strong willingness among stakeholders to embrace the IWRM process irrespective of country, sector and/or occupational background; although the modalities remained fuzzy. The group dynamics observed at the workshops proved that it was possible to bring stakeholders that shared the waters for constructive dialogue, although the political, cultural and institutional context within each of the basins had a major impact on participation. Research project focused stakeholder workshops proved to be a useful tool for enabling soft negotiations on transboundary management of water resources and identifying opportunities for resolving other water use conflicts. It was also noted that projects such as STRIVER could play a "neutral role" in moderating the stakeholder workshops and motivating stakeholders with conflicting interests, by presenting research findings that were perceived to possess a strong heir of legitimacy. The stakeholder workshops also played an important role in offering insights on IWRM practice from other basins around the world and thus promoting awareness and, to some extent, also capacity building. Ultimately, the stakeholder workshops helped in fostering linkages between the STRIVER researchers, managers, end users and policymakers, and at the same time improve acceptance of project outcomes.

Besides the activities mentioned above, the approach also included in the development of a series of Policy and Technical Briefs that were produced in order to communicate the main technical and policy messages to the stakeholders. The content of the Briefs has been presented in a form that the stakeholders can easily understand. Attempts were made to present the Briefs to managers and policy makers in various meetings and workshops. Some of the briefs were translated into local languages to benefit the end users. The main objective was to disseminate the project results to the most relevant stakeholders. In total, 22 Policy Briefs and 13 Technical Briefs have been published. The basin-specific briefs have been presented at the stakeholder meetings and also disseminated at larger events such as the World Water Forum 5 in 2009. The effectiveness of the briefs is yet to be monitored but it is likely that such Briefs can target end-users compared to ordinary research articles and lengthy reports.

3.2 *Environmental Flow Methodology with Science-Stakeholder Interactions*

Another example of how STRIVER has tried to integrate between scientists, non-scientific experts and stakeholders can be found in the work with developing a new environmental flow methodology for hydropower regulation in rivers (Barton et al. 2010; Berge et al. 2010). Since the early 1970s, ecological aspects by the introduction of concepts around minimum or environmental flow (EF) have been given increased attention world-wide, both with respect to assessing the nature of the impacts as well as introducing abatement measures to reduce the negative impacts from hydropower development. EF is defined as "a flow that as far as possible takes care of the entirety and integrity of the ecosystem, the different user interests, and the future resource base in the watercourse" (Halleraker and Harby 2006). There exist more than 200 methods of assessing environmental flows, many of which were too complicated or demanding for use in developing countries (Berge et al. 2010).

A relative simple EF methodology was therefore developed in the STRIVER project. The method entitled "Pressure Impact Multi Criteria Environmental Flow Assessment" (PIMCEFA) seeks to set water release rules that would ensure favourable water levels for river ecology and livelihoods, within the constraints of economic feasibility. The PIMCEFA method does not require comprehensive field investigations such as habitat and aquatic biological registrations and detailed river cross-section descriptions (Berge et al. 2010). According to this method, scientific inputs from river ecologists, hydrologists, environmental scientists, economists are first used to produce the basic knowledge about the functions, values and problems of the river basin. Then an expert panel is established where the scientists, non-scientific experts and stakeholders jointly: (1) define policy-relevant alternatives, (2) identify river ecological and user interests (for which impacts are to be determined), (3) draw optimal water level curves for each ecological value and user interest chosen; and (4) discuss and develop pressure-impact curves for various flow regimes. Finally, the results are translated into a multicriteria tool that pools the information; which can then be discussed with stakeholders in an accessible manner. The methodology itself requires a high level of collaboration with a range of stakeholders and the outcome of the work definitively integrates the knowledge of different expertise and the trade-off of contrasting interests. In the STRIVER project, the methodology was successfully developed in the Glomma and the Sesan river basins (Berge et al. 2010; Barton et al. 2010). It was concluded that Environmental Flow has to be considered as one of the most important measures to reduce negative impacts from hydropower development. In the Sesan River, it is important for preserving aquatic productivity, biodiversity, as well as livelihoods for the local people in downstream reaches (Nesheim et al. 2010a).

3.3 Pollution Modelling with Stakeholders

Accuracy in the quantification of pollution sources and pathways is a major challenge facing the research community as well as managers and policy makers. In the evaluation of environmental changes and management actions river basin models are getting increasingly important. In a science-policy context this implies both selecting the appropriate tools and making sure that the management scenarios have real relevance. In the STRIVER project, these aspects were considered in addressing water pollution related to phosphorus (P) and nitrogen (N) loss from agricultural dominated catchments. Modelling nutrient pollution in Glomma (sub-basins Hunnselva and Lena) and Tungabhadra was carried out using the same methodological approach, i.e. applying the same river basin modelling tool(s) (Grizzetti et al. 2010a). Of major importance was the involvement of local stakeholders at different stages in the modelling process, including the preparation of input related to farming practices, scenario development and analysis. The overall objective of applying models was to obtain information about its suitability to quantify changes in nutrient loss under different management scenarios and whether these changes could meet surface water quality targets. Of major importance was also the evaluation of data availability in the twinned basins, the mutual transfer of know-how (e.g. experiences, concepts, results), technology (e.g., methodologies, models) and modelling procedures. The important role of reliable data such as water quality measurement is indisputable. This is not only a natural scientific dogma but also largely recognized by various international water management bodies (Stålnacke and Gooch 2010). Moreover, it is paramount that modeling tools have to provide reliable results, to be able to provide sound scientific advice to the managers and that economically defensible decisions can be made based on the results. Therefore, the quantification tools have to be accurate in predicting existing nutrient loss and be responsive to changes in land use and land management. A general guideline in the selection of models should be its compliance with the data available. In addition, modellers should preferably have local knowledge about the physical conditions and management practices in the basin, and have a continuous ongoing dialogue with stakeholders to guarantee obtaining reliable scenario results.

Two models were applied, i.e. the TEOTIL (Tjomsland et al. 2009) and SWAT (Neitsch et al. 2002) respectively. SWAT relies on detailed spatial input data, among others soils, cropping systems and climatological data. Less information is needed for the TEOTIL model, which operates mainly on the basis of so-called export coefficients for nutrient loss from different land use types. The overall modelling performance was found to be reasonable. However, at several stages of the modelling process, shortages and gaps in the required data were identified, leading to assumptions and data constructions. Especially in those stages of the modelling process, contact and dialogue with stakeholders was important. Modelling results and management recommendations for the Glomma and Tungabhadra river basins are reported in Grizzetti et al. (2010c) and Lo Porto et al. (2010) respectively.

A participatory modelling approach implies involving stakeholders in model exercises, thereby incorporating local knowledge and understanding of the natural system. In such a process a good communication and interaction between scientist and stakeholders is important to be able to identify and understand the values and motives of a wide range of stakeholders, thereby arriving to logical decisions and management actions. The STRIVER project, interaction and co-operation between scientist and local stakeholders of the respective basins in the development and implementation of scenario modelling was a prerequisite in IWRM (Grizzetti et al. 2010a, b). In the case of the Glomma sub-basins Hunnselva and Lena, measures related to agricultural practices to comply with environmental policy objectives were selected, including among others conservation tillage and optimal fertilization application. Hunnselva is part of Phase 1 in the implementation of the EUs Water Framework Directive (WFD) and as such, the modelling results have been used as supporting knowledge in the planning of mitigation measures in the basin. In the Tungabhadra river basin, the stakeholders were interested in modelling the impacts of climate change, improved sewage treatment, irrigation technology and changed rice production.

The major experience in pollution modelling in the Glomma and Tungabhadra river basins was that stakeholder involvement at different phases of the modelling process, such as input preparation, scenario building and discussions of modelling outcome, played a key role.

3.4 IWRM Assessment in the Four STRIVER Basins

A comparison of IWRM status assessment of the four basins was undertaken (Nesheim et al. 2010a) with a focus on the environmental, socio-economic and institutional dimensions based on the IWRM principles and components suggested by GWP and the 1992 Dublin conference, preparing for the Earth Summit in Rio de Janeiro the same year. In addition, capacity building and transboundary issues were analyzed as they were considered important for the STRIVER project and for implementing IWRM.

The following major conclusions were drawn (Nesheim et al. 2010a):

- Socio-economic development was the common driving force in all the STRIVER river basins, although the role played by each sector varied according to the local conditions. As a consequence, in some cases the river basins face similar pressures and comparable impacts. Negative changes in water quantity and quality were the two common factors in all the conflicts and all the river basins had developed plans or policies to protect most of the environmental aspects. However, it was difficult to assess their level of implementation and the effectiveness of the planned measures. The extent to which problems of water scarcity and water pollution are effectively mitigated depends not only on the existence of relevant law and policy, but also on the degree to which such instruments are implemented. In other words, it is the management and the institutional situation which ultimately determines the outcome. The case basin situations show that there are numerous issues of pollution, environmental flow, water allocation, and more, where laws and policies are not always implemented. This situation may be partly due to weak institutions (particularly in the Sesan and Tungabhadra), but also due to poor coordination between institutions (Nesheim et al. 2010a; Manasi et al. 2010; Gooch and Rieu-Clarke 2010). The latter problem is evident in almost all basins, caused by the multitude of different bodies involved, and unclear definition of roles and functions as well as (in some cases) a lack of financial capacity. In Sesan and Tungabhadra basins, less sufficient monitoring (e.g. crucial water quality parameters) and control and also poor implementation of the prescribed actions were registered.

- Generally it can be argued that in areas where water shortage was evident, measures were taken for water saving and reuse.
- In all the STRIVER basins, access to information and decision-making was present – in particular through environmental impact assessment procedures – at varying degrees. Information laws or the right to information also allowed stakeholders to access information through public authorities. Whilst a number of conventions and national laws and policy provide for stakeholder participation in water management, practice demonstrates that government officials, hydropower, industry and urban interests tend to dominate decision-making (Rieu-Clarke et al. 2010). As a result interest groups representing environmental concerns and communities tend to have less influence in decision-making despite the procedural rules allowing them access to relevant information and procedures, such as public inquiries.
- Capacity building was officially emphasized as part of many water policies and strategy documents in the Sesan, Tungabhadra and Tagus case study basins. However, it was evident that these official statements were seldom operationalised to any larger degree by the authorities. Competence building which includes NGOs are increasingly becoming a part in basins and covers information campaign brochures for the public and training programs for certain sectoral groups (especially for local communities, farmers). In Cambodia, foreign donors were an important actor being responsible for training programs.
- The case basins represent a full spectrum of alternative transboundary contexts/situations, from inter-states (Tungabhadra) to international (Tagus and Sesan). The cases illustrate that the extent of conflict decreases as we move along the continuum from international to local and the extent of conflict varies by type: e.g. competition over the quantity of water was more controversial then conflict over quality. Hydropower development is perhaps the most important issue in a transboundary context, as it may potentially have detrimental effects on the river downstream of the border.
- There exists a considerable number of "IWRM-initiatives", IWRM-plans and policies in all the four basins, but practical full IWRM implementation is generally lacking, except in the case of Glomma (Nesheim et al. 2010b).

3.5 *Land and Water Use Interactions*

In the two river basins, Tagus and Tungabhadra where the comparative study was conducted, land and water use interactions and its implications for IWRM is very closely linked. It is obvious that, and changes in land use exerts pressure on water resources, for e.g., introducing irrigated agriculture in river basins has significantly altered the water allocation patterns and usage, including the water quality. The latter is impacted by use of more chemical fertilizers in irrigated agriculture. On the other hand, changes in water cycle, due to climate change impacts or other local factors, may also lead to changes in land use.

A study by Begueria et al. (2008) has shown a significant land use change in the last decades in both the Tagus and the Tungabhadra basins, due to new water demands. While urbanization has increased in the Spanish part of the Tagus, the increase in natural vegetation in the upstream part of Tagus in Portugal is clearly evident, showing a change in the land use (CORINE Land Cover 1985/1990 and 2000). In the Tungabhadra basin, irrigated agriculture has been the main focus since 80s replacing dryland agriculture, with the construction of the irrigation infrastructure. As a consequence, natural forests and grazing lands have been brought under agriculture. Overall, there has been more urbanization in the peri-urban areas in the Tungabhadra basin.

Although, the changes are not new, the land and water management institutions do not work together. At the basin level, there is a lack of an authority that can integrate the management initiatives. Integrating the efforts of various relevant sectors, could provide a better management option in both Tagus and Tungabhadra, where the challenges appear more complex, and cannot be addressed by sectoral approach. IWRM can be a practical tool in both the basins to resolve transboundary issues and water conflicts between sectors. This may be feasible at least at a sub-basin level to start with as a management unit, if properly planned. In the two basins, we have observed that water management and planning, does not adequately take into consideration the linkages between land and water use.

There are several options, as analyzed by our teams in the STRIVER project that can be recommended, such as improving the efficiency of water use, reducing competition between sectors and introducing new cropping systems and practices that would need less water than the current systems. For example, crop rotation between irrigated and dry land crops, new irrigation methods like drip and sprinkler systems that increases the productivity, introducing inland fisheries and aquaculture, and water and soil moisture conservation measures. Farming and small-scale fishery is predominant livelihood source in many developing countries. However the latter group is a marginal group that is not integrated with activities of other sectors, for example water for irrigation or considered when Water Resources Department make water allocation priorities (Joy et al. 2010).

We have observed new initiatives in the two basins, and efforts to move towards integrated management. However, some of these efforts are still not in practice, and in fact, the stakeholders have not been prepared for such a change. The change has to come from bottom up, with a top down institutional supportive mechanisms.

There is some willingness to change in both the basins, as observed from the new policy documents, new programs attempting to bring the relevant stakeholders at the planning phase and capacity building initiatives. At least in the Tagus, the governments are bound by the EU Water Framework Directive and are required to comply with some of the directive regulations that require integrated management. In Tungabhadra, no such directives exist. It is still a long way to go in both the Tagus and Tungabhadra basins before we can claim that IWRM is in practice and stakeholders are actively involved in managing water resources.

3.6 Transboundary Conflicts and Water Governance

Transboundary settings are especially sensitive in IWRM, as conflicts over the use and sharing of water and the demands placed on water governance increase in transboundary waters (Gooch and Stålnacke 2006). Different national and sectoral interests from different countries must try to cooperate in the use of transboundary water and while water laws and systems of administration are mostly unified in national contexts, in international rivers, there are with a few exceptions no unitary authorities that can legally or politically force actors in the water sector to comply to laws and agreements. Competition between different agencies does of course also exist in national contexts, where different agencies and departments often compete over resources, funding and information. The special conditions involved in sharing a common resource such as water between different countries also creates conditions that are especially dependent on efficient legal systems and communication. The challenge is that these often do not exist, which makes cooperation between the different organizations and institutions, such as governmental agencies and departments, more complicated. Transboundary conditions can also result in challenges for NGO's and other stakeholder groups to create efficient means of communication and influence with these groups of policy-makers and managers (Gooch and Rieu-Clarke 2010). Transparent water management and the involvement of stakeholders and civil society need functioning channels of communication through a number of networks which may consist of official and unofficial actors, formal and informal institutions. These networks may include different official departments and agencies, funding agencies, NGO's, village and community informal organisations etc., as well as infrastructures in the river basin such as hydroelectric power plant dams (Gooch 2008a). Within these networks, the world-views and understanding of the actors and groups of actors strongly influences their treatment of information and knowledge. The scientific information provided by a project such as STRIVER may be accepted or rejected according to these mind-frames (Gooch 2008b).

While a significant body of research has sought to examine governance in general, little work has been done to tailor such work to the specific context of (transboundary) IWRM. The STRIVER work related to governance therefore sought to develop a robust set of indicators, capable of assessing laws, policies and institutions related to (transboundary) IWRM. An important first phase of the work was to review the applicability of existing theory related to governance, IWRM and indicator analysis (Rieu-Clarke et al. 2008; Allan and Rieu-Clarke 2010). This research was the basis by which STRIVER developed a set of indicator questions. The question focused on examining the extent to which good governance principles, e.g., accountability, transparency, participation and predictability, were embedded within the laws at multiple-scales (international, regional, national and provincial). The work also analysed the degree to which good governance principles – enshrined within the law – were translated into practice. Clearly such an examination required extensive empirical analysis, which was secured through collaboration with stakeholders via workshops and interviews. Such empirical analysis was vital in order to ascertain the extent to which laws were implemented. Stakeholder engagements also offered an important means by which to "validate" the research outputs from the project. An underlying theme running throughout the research findings related to governance was the need to account for, and fully understand, the wider governance context in which decision around IWRM are taken (Rieu-Clarke et al. 2008; Allan and Rieu-Clarke 2010).

4 Concluding Remarks

4.1 STRIVER Findings in an Overall IWRM Context

We have claimed in this paper that IWRM seeks to achieve a balance between economic efficiency, social equity and environmental sustainability (see e.g., Gooch et al. 2010). As pointed out by Gooch and Stålnacke (2006), the definition of IWRM and what is usually understood as the main contents of the ideas behind it, have a strong focus on the concept of sustainability,, yet little is, however, provided as to how this integration, balancing of interests and co-ordination is to be achieved. The practical implementation of IWRM, its operationalisation, and how to achieve measurable criteria of its success or failure, have also been questioned (e.g., Jeffrey and Gearey 2006; Lankford and Cour 2005; Biswas 2004). Others have claimed that the benefits and added-values of the idea must be clearly shown in order to achieve political and public acceptance for IWRM (Ballweber 2006). The need within IWRM to include the full range of physical, biological, and socioeconomic variables has also been stressed as central to the IWRM process (Hooper 2003). Unfortunately, such a knowledge base is usually lacking or not accessible, or it is scattered across various sectoral agencies which may be unwilling or unable to cooperate with one another. Other fundamental attributes usually considered part of IWRM include social and motivational feasibilities (Chess and Gibson 2001). The important role of science has lately also been emphasised (Gooch and

Stålnacke 2010a; Quevauviller 2010). More specifically, it has been claimed that scientists can play a key role as a neutral third party (Gooch and Stålnacke 2010a), especially in local capacity building (Breen et al. 2004).

What then are the main findings of the STRIVER project and how can these findings help us to improve intergrated water management? Stålnacke et al. (2010) have formulated the following main findings from the STRIVER project:

- Stakeholder participation is one of the most critical elements for practical IWRM implementation. It helps not only in the integration of various perspectives of stakeholders from different sectors, but also of different user groups with varied and conflicting interests.
- Research projects such as STRIVER can act as independent facilitators and provide a neutral platform for stakeholder dialogue, which ultimately can help facilitate the IWRM process. During the project life there was seen a relatively strong willingness among stakeholders to participate in the IWRM process during the workshops, irrespective of their country, sector and occupational background. The stakeholder workshops were also seen to play an important role in capacity development initiatives.
- Cross-border cooperation was a difficult process and sensitive to address. In all the three transboundary river basins the challenges of cooperation were seen as a potential hindrance to IWRM and therefore needed to be carefully addressed. It was only in the stakeholder workshops that STRIVER could bring together actors from different countries or provinces sharing the river waters.
- The development and show-casing of various "tools" (e.g., environmental flow, pollution models, water pricing, Actor-Network analyses and scenarios) is of high interest and considered to be of high value for the water managers. One common feature with all the applied and tested tools used in the project were that they helped promote dialogue and integration between the different actors involved in IWRM as well as between scientists and stakeholders.
- There exist a considerable number of "IWRM-initiatives", IWRM-plans and policies in all the four basins, but practical IWRM implementation is generally lacking, except in the case of Glomma.

4.2 Some Reflections on How to Strengthen European Research in the Perspective of Transformative Research for Sustainable Development

The challenge facing European research is not one of scientific quality, nor one of funding, but of funding and research aimed at interaction with managers, end-users and policy-makers. Despite a number of significant efforts on the part of the European Commission, much of the output of EU funded projects is still not used by the very people who would have most use for it (Quevauviller 2010). Why? Many R&D projects simply do not have the flexibility, or the will, to embark on the process of learning to communicate their results to people who are not experts in their field of study. This is partly a result of the constraints of EU funding, and partly the result of embedded academic traditions that prioritise internal presentations of results in scientific journals, many of them only read by a limited number of people. Another problem in ordinary research projects is that the objectives need to be defined ex ante (Jäger 2011). Compared to many other R&D projects, twinning projects, like STRIVER, did however had the flexibility to change the research focus during the duration of the projects and also interact with water managers and policy makers (see e.g., Sect. 3.1 above). However, this interaction is dependent on scientists taking the role upon themselves of "scientific ambassadors". In addition, it also demands more time and resources in terms of logistics required for travels, meetings and discussion forums. However, this in itself is not enough; as noted in other sections of this chapter what is needed is the development of an interface, not one-way communication. In other words: the integration of scientific research with policy, management and society has to be considered at all stages of IWRM development, including design, planning, implementation and review (Grizzetti et al. 2010b). Managers and policy-makers must also be prepared to take the time and make the effort to interact with the research community. Managers and policy-makers can be initially are sceptic to "spend" their valuable time with researchers; we observed this in the initial stages of the STRIVER project. In the course of time, STRIVER managed however to establish trust and confidence amongst the stakeholders, and to develop ways to interact with these groups and to integrate their knowledge into the project. Unfortunately, for researchers, these kinds of stakeholder interactions can often only be achieved and developed at the expense of scientific publications, and it is these publications that are the base of a scientific career. If results from projects such as STRIVER are to reach managers and policy makers, it is necessary to spend the resources and to establish formal links with stakeholders (from local level, to high-level water managers and policy makers) and the research community. This is particularly relevant to the implementation of IWRM and other sustainability programmes, and this is something that must be developed during the whole course of the project. Such a link could be win–win situation for both researchers and managers, and also help in a better understanding of the research problems to be addressed It could also lead to the joint development of appropriate tools, scenarios and policy guidelines. Another important result of the STRIVER project that was experienced during the field trips and workshops was that stakeholders were interested in capacity building that could be also formalized in the projects. In more practical terms, it is therefore recommended that funding agencies like DG Research at the EC continue with the twinning mechanisms in the future, but with more emphasis on twinning and interaction between the scientific community, stakeholders and the public. This would involve, in order to be effective, projects with at least 4–5 years project duration. The absence of long-term funding to support iterative and participatory processes for sustainability research projects is also emphasised and further discussed by Jäger (2011).

Acknowledgements STRIVER: Strategy and methodology for improved IWRM – An integrated interdisciplinary assessment in four twinning river basins (2006–2009) is a project supported by the European Commission Sixth Framework Programme (FP6) under SUSTDEV-2005-3.II.3.6: Twinning European/third countries river basins. Contract number: 037141.

References

Alcamo, J. (2001). *Scenarios as tools for international environmental assessments* (Report No. 31). Copenhagen: European Environment Agency.

Allan, A., & Rieu-Clarke, A. (2010). Good governance and IWRM: A legal perspective. *Irrigation and Drainage Systems, 24*, 239–248.

Ballweber, J. A. (2006). A comparison of IWRM frameworks: The United States and South Africa. *Journal of Contemporary Water Research & Education, 135*, 74–79.

Barton, D. N., Berge, D., & Janssen, R. (2010). Pressure-impact multi-criteria environmental flow analysis: Application in the Øyeren delta, Glomma River basin, Norway. In G. D. Gooch & P. Stålnacke (Eds.), *Integrated transboundary water management in theory and practice: Experiences from the New EU Eastern border* (pp. 35–48). London: IWA Publishing.

Beguería, S., Vicente-Serrano, S., López-Moreno, I., Lana-Renault, N., & García-Ruiz, J.M. (2008). Land use change and water resources in the Tagus and Tungabhadra basins. STRIVER Technical Brief No. 1 (http://www.striver.no/diss_res/files/STRIVER_TB1.pdf).

Berge, D., Barton, D. N., Nhung, D. K., & Nesheim, I. (2010). The science-policy-stakeholder interface and environmental flow. In G. Gooch & P. Stålnacke (Eds.), *Science, policy and stakeholders in water management* (pp. 105–122). London: Earthscan.

Biswas, A. K. (2004). Integrated water resources management: A reassessment. *Water International, 29*(2), 248–256.

Breen, C. M., Jaganyi, J. J., van Wilgen, B. W., & van Wyk, E. (2004). Research projects and capacity building. *Water SA, 30*(4), 429–434.

Chess, C., & Gibson, G. (2001). Watersheds are not equal: Exploring the feasibility of watershed management. *Journal of the American Water Resources Association, 37*(4), 775–782.

Gooch, G. D. (2008a) Actor-Network Theory in water management – a help or hinder for understanding water management regimes? STRIVER Technical Brief No. 2. (http://www.striver.no/diss_res/files/STRIVER_TB2.pdf).

Gooch, G. D. (2008b) Communication in IWRM in transboundary rivers. STRIVER Policy Brief No. 6. (http://www.striver.no/diss_res/files/STRIVER_PB6.pdf).

Gooch, G. D., & Huitema, D. (2008). Participation in water management: Theory and practice. In J. G. Timmerman, C. Pahl-Wostl, & J. Möltgen (Eds.), *The adaptiveness of IWRM – analyzing European IWRM research*. London: IWA Publishing.

Gooch, G. D., & Rieu-Clarke, A. (2010). IWRM status in the Sesan River Basin. In G. D. Gooch, A. Rieu-Clarke, & P. Stålnacke (Eds.), *Integrating water resources management: Integrating methodologies and strategies in practice* (pp. 128–140). London: IWA Publishing.

Gooch, G. D., & Stålnacke, P. (Eds.). (2006). *Integrated transboundary water management in theory and practice: Experiences from the New EU Eastern border*. London: IWA Publishing.

Gooch, G. D., & Stålnacke, P. (2010a). *Science, policy and stakeholders in water management* (p. 166). London: Earthscan.

Gooch, G. D., & Stålnacke, P. (2010b). Introduction: The science-policy-stakeholder interface. In G. Gooch & P. Stålnacke (Eds.), *Science, policy and stakeholders in water management* (pp. 1–15). London: Earthscan.

Gooch, G. D., Rieu-Clarke, A., & Stålnacke, P. (2010). *Integrating water resources management* (p. 160). London: IWA Publishing.

Grizzetti, B., Lo Porto, A., Barkved, L. J., Joy, K. J., Paranjape, S., Deelstra, J., et al. (2010a). The science-policy-stakeholder interface in water pollution assessment. In G. Gooch & P. Stålnacke (Eds.), *Science, policy and stakeholders in water management* (pp. 67–82). London: Earthscan.

Grizzetti, B., Bouraoui, F., Gooch, G. D., & Stålnacke, P. (2010b). Putting the 'integration' in the science-policy-stakeholder interface. In G. Gooch & P. Stålnacke (Eds.), *Science, policy and stakeholders in water management* (pp. 17–28). London: Earthscan.

Grizzetti, B., Bouraoui, F., Barkved, L. J., & Deelstra, J. (2010c). Modeling water nutrient pollution with stakeholders. In G. D. Gooch, A. Rieu-Clarke, & P. Stålnacke (Eds.), *Integrating water resources management: Integrating methodologies and strategies in practice* (pp. 25–33). London: IWA Publishing.

GWP-TEC (Global Water Partnership – Technical Advisory Committee). (2000). *Integrated water resources management* (TAC Background Papers No. 4). Stockholm: GWP.

Halleraker, J. H., & Harby, A. (2006). *International methods for deciding environmental flow which of these are aplicable in Norway? NVE Miljøbasert vannføring* (Report 9 2006, pp. 69). ISBN 82-410-0584-9.

Hooper, B. P. (2003). Integrated water resources management and river basin governance. *Water Resources Update, 126*, 12–20.

Jäger, J. (2011). Risks and opportunities for sustainability science in Europe. In C. C. Jaeger, J. D. Tàbara, & J. Jaeger (Eds.), *Transformative research for sustainable development*. New York: Springer, pages 185–201.

Jeffrey, P., & Gearey, M. (2006). Integrated water resources management: Lost on the road from ambition to realisation? *Water Science and Technology, 53*, 1–8.

Joy, K. J., Paranjape, S., Manasi, S., Mollinga, P., & Nagothu, U. S. (2010). Tungabhadra sub-basin: Recommendations. In G. D. Gooch, A. Rieu-Clarke, & P. Stålnacke (Eds.), *Integrating water resources management* (pp. 121–126). London: IWA Publishing.

Lankford, B. A. & Cour J. (2005, March 7–9). *From integrated to adaptive: A new framework for water resources management of river basins*. In the Proceedings of the East Africa River Basin Management Conference, Morogoro, Tanzania.

Lo Porto, A., De Girolamo, A. M., Gosain, A., & Barkved, L. J. (2010). Water quality assessment and water pollution modelling in the Tungabhadra river basin. In G. D. Gooch, A. Rieu-Clarke, & P. Stålnacke (Eds.), *Integrating water resources management: Integrating methodologies and strategies in practice* (pp. 105–119). London: IWA Publishing.

Manasi, S., Nesheim, I., Joy, K. J., Paranjape, S., Nagothu, U. S., Raju, K. V., et al. (2010). IWRM status in the Tungabhadra river basin. In G. D. Gooch, A. Rieu-Clarke, & P. Stålnacke (Eds.), *Integrating water resources management: Integrating methodologies and strategies in practice* (pp. 83–91). London: IWA Publishing.

Neitsch, S. L., Arnold, J. G., Kiniry, J. R., Williams, J. R., & King, K. W. (2002). Soil water assessment tool theoretical documentation. Temple Texas: Grassland, soil and water research laboratory. Agricultural Research Service, Blackland Research Center, Texas Agricultural Experimental Station.

Nesheim, I., McNeill, M., Joy, K. J., Manasi, S., Nhung, D. T. K., & Manuela, M. (2010a). The challenge and status of IWRM in four river basins in Europe and Asia. *Irrigation and Drainage Systems, 24*(3–4), 205–221.

Nesheim, I., Stålnacke, P., Nagothu, U. S., Skarbøvik, E., Barkved, L. J., & Thaulow, H. (2010b). IWRM status in the Glomma river basin, 2010. In G. D. Gooch, A. Rieu-Clarke, & P. Stålnacke (Eds.), *Integrating water resources management: Integrating methodologies and strategies in practice* (pp. 13–23). London: IWA Publishing.

Quevauviller, P. (2010). Is IWRM achievable in practice? Attempts to break disciplinary and sectoral walls through a science-policy interfacing framework in the context of the EU Framework Directive. *Irrig Drainage Syst* (2010) 24:177–189. DOI 10.1007/s10795-010-9102-x.

Rahaman, M., & Varis, O. (2005). IWRM: Evolution, prospects and future challenges. *Sustainability: Science, Practice, and Policy, 1*(1), 15–21. doi:URL: http://ejournal.nbii.org.

Rieu-Clarke, A., Allan, A., & Magsig, B. O. (2008). *Assessing governance in the context of IWRM* (STRIVER Policy Brief No. 8). (http://www.striver.no/diss_res/files/STRIVER_PB8.pdf).
Rieu-Clarke, A., Baggett, S., Campbell, D., Joy, K. J., & Paranjape, S. (2010). The science-policy-stakeholder interface and stakeholder participation. In G. Gooch & P. Stålnacke (Eds.), *Science, policy and stakeholders in water management* (pp. 29–50). London: Earthscan.
Shell International. (2003). *Exploring the future. Scenarios: An explorers guide*. London: Shell International Limited.
Stålnacke, P., & Gooch, G. D. (2010). Integrated water resources management. *Irrigation and Drainage Systems, 24*, 223–238.
Stålnacke, P., Gooch, G. D., & Rieu-Clarke, A. (2010). STRIVER – Overall findings. In G. D. Gooch, A. Rieu-Clarke, & P. Stålnacke (Eds.), *Integrating water resources management: Integrating methodologies and strategies in practice* (pp. 151–160). London: IWA Publishing.
Tjomsland, T., Selvik, J. R., & Brænden, R. (2009). *Teotil. Model for calculation of source dependent loads in river basins* (p. 49). Oslo: NIVA.

Pragmatism and Pluralism: Creating Clumsy and Context-Specific Approaches to Sustainability Science

Paul M. Weaver

1 Introduction

In the words of Tim O'Riordan, sustainable development is a 'tough nut to crack' because it does not fit easily with the normal political model of analysis and decision. The urgent need for sustainable development is evident, but the concept is vague, contradictory and confusing. O'Riordan points out that there is no agreement on what sustainability actually is, where we have to go to get it, and what it would look like in a multi-national world of nine plus billion people demanding more and more from a stripped and stressed planet (O'Riordan 2008). O'Riordan is right in that it is difficult to pin down what sustainable development is and what sustainability transitions will imply, for reasons which include that sustainability transitions are dynamic, systemic, configuration dependent, and indeterminate.

But whereas it may be difficult to pin down sustainability it is not difficult to pin down the mechanisms at play that make current pathways unsustainable and, from this, we can make further progress. We can highlight directionally-reliable trends for sustainability transitions to steer toward and push forward. We can obtain insights into steps that could be taken top-down to create market and regulatory framing conditions that would improve prospects for sustainability transitions generally, especially by increasing the degrees of freedom available for context-specific initiatives and innovations to emerge from bottom-up. Equally we can look at niche examples and contexts where transitions to apparently more sustainable ways of exploiting ecosystems and of creating sustainable livelihoods and wellbeing for community members are underway already. Importantly for the theme of this book and for those engaged in sustainability science we can also draw inferences about the roles science might play in supporting sustainability transitions. This is the core concern of the present chapter.

P.M. Weaver (✉)
Groundswell Research Associates
e-mail: pweaver.groundswell@gmail.com

C.C. Jaeger et al. (eds.), *European Research on Sustainable Development*, 173
DOI 10.1007/978-3-642-19202-9_12, © European Union, 2011
Published by Springer-Verlag Berlin Heidelberg 2011

Clarifying the roles of science in sustainability is now an urgent issue. It may even be 'defining' for the prospects of sustainability transition. What happens next – that is whether effective progress on sustainability transition will be made – is on the cusp. Decisions and actions taken in the next 5–10 years will be profoundly important for the progress that is possible and whether the most serious consequences of unsustainable development can be averted, especially as concerns communities whose livelihoods and wellbeing are most vulnerable to environmental change threats and who are most marginalised by powerful macro-scale forces.

2 Purism Versus Pragmatism

In seeking to clarify what roles science could usefully and most effectively play in sustainability transitions, this chapter seeks to distinguish between 'purist' and 'pragmatic' stances. Purist stances stress the importance of a particular approach as being the 'key' for sustainability. These often stress a particular structural theme such as policies, markets or technologies and advocate typically whole-scale reforms to prevailing conditions on the themes concerned. Purism reflects particular stances in respect to the 'how' question of what might be done in a consistent, 'top-down' logic to chart future development pathways. Pragmatism, by contract, stresses that change processes in complex systems are multi-level, multi-dimensional and multi-speed and that innovations developed at lower levels of scale need to be adapted both to local contexts and to higher-level framing conditions and dynamics.[1] Pragmatism suggests that sustainability at the global level depends on achieving a patchwork of sustainable solutions at lower scale levels. In turn this stresses the need for combinations of mutually supportive and reinforcing top-down and bottom-up innovations that engage synergies among innovations on many different fronts.[2]

Science could play a critical role in delivering sustainability transitions. But the deployment of science and scientists in this endeavour to date has been ineffective (sometimes profoundly so) and may perhaps even have been counter-productive in some aspects. This will continue to be the case until the challenge that sustainable development represents to society and to scientists is better and more widely understood and, in turn, this is translated into new mandates for scientists and

[1]Other authors in this volume also point to the inherent paradoxes and contradictions of sustainable development, which complicate the process of achieving consensus and of establishing directions that are purist, unambiguous and non-controversial. To factor into the mix, also, is that pathways to sustainability need to be elaborated in context. Pluralism, diversity and context-specificity also are acknowledged leitmotivs for sustainable development.

[2] The analogy might be drawn in the contrast between 'purist' scenarios that scientists construct to help explore how different futures could unfold and which are typically delineated along the lines of dominant or leading themes and 'hybrid' scenarios that explore interactions among innovations of many different types. See, for example, the 'market first', 'security first', 'policy first' and 'sustainability first' scenarios developed for the Global Environment Outlook, UNEP 2007.

researchers. The roles such mandates specify for scientists are different from conventional and 'purist' roles and they call for a distinctive new form of science for sustainable development. We begin therefore by reflecting on key insights learned about sustainable development over the years since the concept first came to prominence, so that we might be able to better specify the challenge that sustainable development presents both to society at large and to those in the scientific and research communities who seek to support societal processes of sustainability transition.[3]

3 An Urgent Challenge

Making effective progress on sustainability transition is urgent and has been made so because each year there is an absolute increment in anthropogenic environmental stress. The global economy has become so large that even small rates of economic growth translate into large annual increments in absolute levels of economic activity, metabolic 'throughput', and stress. This absolute growth in the demands the global economy makes for environmental resources and ecosystem services is rapidly closing down any remaining distance to resource and ecological safety thresholds and is accelerating us toward these at the same time as these are 'closing in' because of anthropogenic environmental change. By contrast, no systematic progress is being made on poverty reduction in absolute terms.

Research over the last two decades has documented that the Earth is undergoing major environmental and socioeconomic changes (e.g. Steffen et al. 2004; UNEP 2007). Climate change, land degradation, deforestation, biodiversity loss, and changes of water quality and quantity are prominent examples of global environmental changes. Globalisation, demographic changes and the widening gap between rich and poor are examples of socioeconomic trends that are linked closely with such environmental changes. Recently a group of scientists proposed nine ecological boundaries or safety thresholds to preserve planetary mechanisms and environmental conditions on which humanity and much of the rest of the biosphere depends. They advised that these should not be crossed as this risks generating unacceptable environmental change for humanity (Rockström et al. 2009). Three of these may have been crossed already.

[3]This includes scientists and researchers, but also those who set research priorities and agendas, those who allocate and administer funds for research, and those who establish and implement the criteria used to evaluate research proposals and research outcomes.

4 Conventional Diagnoses

It is unsurprising, therefore, that one of the conventional diagnoses holds that the core challenge of sustainability is one of increasing the eco-efficiency with which goods and services are produced and consumed so that economic growth might be 'decoupled' from environmental stress. This diagnosis is unproblematic for the mainstream. It fits the prevailing development paradigm that equates progress with economic growth. It appeals to conventional political and business logics since it seemingly provides a 'way out' of the dilemma of ecological limits to growth. And it indicates a broadly conventional role for scientific work in support of sustainable development, for example in undertaking technical research into resource substitutions, developing eco-efficient technologies, increasing resource and energy productivities, and reducing emissions of wastes and pollution.

It is equally unsurprising that economic growth has historically been the dominant goal of modern societies, since the position from which modernisation began in most societies was typically one of material deprivation and hardship for the majority. More interesting in the current context is to understand why, as the richer societies have modernised and, one-by-one, have overcome material shortfalls and extended affluence across their populations, they continue to pursue further economic growth, even in the knowledge that this is raising ecological liabilities. This is largely because of mechanisms that are in-built in the way a modern market economy operates. These generate *either* spirals of growth *or* spirals of recession, but are unable easily to establish a 'steady-state'.

A market economy works by harnessing competition, which stimulates investment and innovation in a never-stopping search to increase returns to capital. Through innovation there is a continuous effort to improve the efficiency of using different factors of production. Innovation efforts focus on increasing the productivity of the more costly factors, since this increases total factor productivity and overall return to capital most effectively. On this basis, competition in the market achieves both increases in natural resource productivity and in labour productivity, but the dominant focus of innovation in markets as they are shaped now is to increase labour productivity, since labour is both relatively expensive as a factor of production and is a more problematic factor to manage and control relative to other factors, such as machinery, raw materials and energy.

The downside of increasing labour efficiency is that fewer workers are needed to produce the same output. Their unemployment risks reducing effective demand for goods and services in the economy. Sustaining social and economic stability – avoiding a recessionary spiral – therefore becomes a core concern of governments and businesses in market economies. Some stimulus to aggregate demand is provided by businesses passing on part of the cost reduction achieved by productivity improvement in the form of lower prices to consumers. Lower prices make goods more affordable and this can create a self-reinforcing cycle and engine for growth so long as the increase in aggregate demand encouraged by price reductions is sufficient to retain labour that would otherwise be released. Aggregate demand may be stimulated or supported also by public spending or by increasing liquidity by providing consumers with easy access to credit. To limit the public debt burden of stimulating aggregate demand through public spending and deficit financing, some governments have preferred to use monetarist policies, which encourage and facilitate private debt as an alternative to public debt as a means to maintain consumption growth. The roots of the recent financial crisis lie in concerted effort to free up credit across the globe to support economic expansion (Jackson 2009). This is one reason why ecological and financial aspects of unsustainable development are systemically interlinked.

5 Self-defeating 'Decoupling'

For a market economy that is growth-oriented, further growth in the economy on a continuing basis therefore becomes necessary to secure economic, social, financial and political stability. This goes a good part of the way to explaining why growth is pursued as a conscious policy goal in its own right irrespective of whether production and consumption of goods and services at the margin is warranted in net benefit terms. It explains why there is political reticence to integrate resource depletion, pollution, and ecosystem degradation costs fully into the market, as well as social costs, since integration of these would make goods and services more expensive, reduce consumer demand, and risk recession. And it explains why so much emphasis has been placed on the strategy of 'decoupling'; i.e. increasing resource productivity as a way to 'decouple' economic growth from the material throughput of the economy. Intuitively decoupling appears to offer scope to reconcile the conflicting needs to expand growth (needed indefinitely to secure economic, financial and social stability as markets are formed now) while containing the physical scale of the economy (needed to ensure planetary boundaries are respected and that the economy stays within ecological safety zones).

Great store has been placed on decoupling. But it is clear from the discussion above that decoupling is not a plausible strategy of itself to deliver sustainability transition. Certainly to reduce the throughout of the economy there is a need to make improvements in resource efficiency and to reduce the pollution intensity of goods and services per unit of GDP. Relative decoupling of GDP from throughput is a necessary condition for ecological sustainability. But it is not a sufficient condition. For the throughput of the global economy to be reduced absolutely, which is what matters for global ecological sustainability, the rate of eco-efficiency improvement must be large enough annually to offset the combined impact of growth in population and growth in average income. It would be needed also for the eco-efficiency gain to be 'captured' and 'dedicated' to reducing the absolute throughput of the economy, rather than being redeployed to support economic growth. Yet the market is structured and oriented currently in a way that ensures that gains in efficiency to all factors of production are dedicated to expanding consumer demand.

Under present market arrangements investment in research and technology development to accelerate eco-efficiency is therefore unlikely to translate into absolute decoupling. Counter intuitively, to the extent that innovation efforts are successful in delivering specific efficiency and productivity improvements, such as reductions in CO_2 emission intensity at the level of specific products and services, this may even make matters worse, not better. Neither is it safe to rely on the argument that increasing affluence will ultimately provide an automatic mechanism for societies to switch away from economic growth. As markets are structured currently growth has its own dynamic unrelated to any material benefits of production and consumption. There is no reason (or credible evidence) to expect a production and consumption 'ceiling' above a certain level of affluence or for absolute decoupling to occur automatically once this level has been attained. In short, there is little likelihood under current market arrangements to 'grow out' of the problem or to assume the problem can be solved with a purely technical fix. Absolute decoupling at the global scale is a highly improbable outcome as markets are currently structured.

6 The Significance of the Market

Important here is the importance of the market and its construction in terms of global economic, social and environmental dynamics and outcomes. The market constitutes the dominant mechanism for strategic coordination at global scale. It is unrivalled in this role. Effectively the market operates de facto as *the mechanism of global governance*. How the market is structured and formed and how this orientates and harnesses key forces and directs resources is critical in shaping economic, social and ecological futures. As the market is formed and operates currently, the relationships between market processes and sustainable development are ambiguous at best. Some progress is being made on increasing relative eco-efficiency, but the rates of progress are below those needed and the productivity gains are not being captured to reduce absolute levels of environmental stress. Environmental change is already upon us and therefore cannot be ignored. Equally, although levels of economic output continue to expand, the gap between rich and poor is widening, not closing.

None of this is to argue against markets or globalisation processes. To the contrary there is a strong case to be made for making progress toward (re)forming markets and globalisation processes so that their powerful capacities for stimulating and coordinating innovation are directed and harnessed toward a widespread restructuring of production-consumption systems and a more sustainable exploitation and protection of critical ecosystems. The relationships between market processes and sustainable development are complex, ambiguous, and configuration-dependent, so a more nuanced understanding of important constituent mechanisms and drivers of market processes and a better differentiation of widely used terms, such as growth, globalisation, investment, innovation, and productivity, would help reveal how the processes might be made more compatible. Current uses of these terms are ambiguous because they conflate ends and means and fail to distinguish adequately between different contexts and targets.

Growth is still needed in poorer countries to overcome poverty. By contrast, a shift away from further material growth in the already wealthy countries would help release environmental space for growth elsewhere. Global savings still need to be marshalled worldwide, but for the purpose of directing these into ecological investments of various forms rather than of using them as sources of cheap and unsecured credit to bolster consumer spending in the already- richer countries. Innovation is needed to increase eco-efficiency massively, but frameworks are needed to enable the gains to be captured to secure absolute reductions in the throughput of the global economy. By contrast, innovation to improve labour productivity is likely to be counterproductive in many contexts, especially where this causes social and livelihood insecurity needlessly. Greater precision and differentiation when using such terms would have important implications for policy, especially concerning the longer-term scope for introducing market reforms that could help to better align market processes with sustainable development targets.

Equally, the global-scale nature of environmental change and related economic, technological and social change processes means that the issues involved cannot be addressed effectively without cooperation at the global scale and without deliberate efforts to coordinate between local- and global- scale responses and between richer and poorer countries and communities. This is easy to demonstrate. In the case of climate stabilization, for example, it is now acknowledged that the richer countries of the OECD acting alone could not put the world onto a 450 ppm trajectory, even if they were to reduce their own CO_2-eq emissions to zero (International Energy Agency 2008). In the same vein there is an inherent symmetry in the approaches needed to manage and protect global-scale and local-scale common pool resources. Decisions by the international community to try to protect globally- and internationally- significant common pool resources, such as the atmosphere, critical habitats and biodiversity, and to develop supporting mechanisms including more extensive use of PES (payments for ecosystem services) schemes, depend for effective implementation on how local communities in regions such as Latin America manage and exploit critical local ecosystems, such as forests, mangroves and coral reefs, which are also, often, common pool resources.

Against these insights a broader debate about sustainable development is now opening up. A starting point has been to revisit the question of how wellbeing, happiness and prosperity are constituted and of the roles of production, consumption and formal economic activity in their delivery. Clearly wellbeing has a material dimension, relating to the fulfilment of basic needs for food, water, shelter and clothing, but it also has psychological and social dimensions, relating to whatever contributes to giving meaning and purpose in life. Wellbeing and happiness can be produced and delivered in many very different ways: via goods and services produced through formal economic activities, through direct ecosystem services, through services provided by public infrastructures, through relationships between and among individuals and society, etc. There are different degrees of substitutability among different sources and forms of wellbeing. Basic material needs are not easily substitutable whereas psychological and social wellbeing may be fulfilled in very different ways. Some sources of wellbeing may be damaged or degraded in producing others. The situation is therefore complex and dynamic. Empirical studies reveal that at low levels of income per capita, increases in income contribute substantially to improving life quality, wellbeing and happiness. However, this direct correlation levels off at around $15,000 per capita. Growth in per capita income above this level does not appear to add significantly to wellbeing or happiness (Worldwatch Institute 2008).

7 Sustainable Development as a Governance Challenge

This background suggests that the core challenge to science of supporting sustainability transition is not primarily a technical or a technological challenge of increasing eco-efficiency *per se*, albeit that this is one of several important components of an overall approach to delivering more sustainable development. Rather, the core challenge is one of governance of sustainability transitions. The top-down need is to ensure over the longer-term that the market and globalisation processes that are the dominant mechanisms of coordination at the global scale are reformed gradually to harness their powerful forces to the achievement of sustainability targets: to ensure that the basic needs of all are met; to stimulate savings and to marshal these into eco-investments of different forms; and to direct achieved improvements in eco-efficiency toward reducing the absolute metabolic throughput of the global economy. There are many candidate macro-scale market reforms that could help here: internationally agreeing and implementing effective planetary safety standards; creating markets for ecosystem services; full cost pricing; and extending producer liabilities on goods entering trade, for example.

But there are important pragmatic considerations to take into account concerning the likelihood and feasibility of introducing different kinds of policy interventions and at different levels and the pace at which interventions can be introduced. Changes in macro-scale market conditions will most likely be introduced only gradually and through phased interventions. Interventions that address particular issues or sectors or are generated and implemented at the spatial scale of particular countries or blocs are likely to be more feasible in the short term. But even modest policy interventions – if well chosen, well designed and well targeted – could critically change the framing conditions for lower-level innovations and work synergistically alongside other (non-policy) sources of change to provide a continually-changing innovation context that opens up new degrees of freedom, new opportunities, and scope for ‘clumsy’ and ‘context-specific’ solutions to emerge from bottom-up.

The twin problems of environmental change and poverty are upon us here and now and they call for action at lower levels of scale where the problems are being felt. This reinforces the point just made that there is an important new role for science in supporting local communities in improving their own situation in respect to their own targets and goals through actions they are able to take themselves in their own contexts of living and operating, taking advantage wherever possible of opportunities created by the higher-level dynamics in regulatory and market frameworks. Important progress is being made already in some policy areas that impinge significantly on market conditions and contexts. The politics of climate change, for example, and associated innovations, such as REDD (Reducing Emissions from Deforestation and Forest Degradation), are increasing the scope for creative and innovative bottom-up approaches to emerge in many Asian, African and Latin American countries. Such initiatives will shortly be reinforced by others developed to support implementation of the Convention on Biological Diversity. There is therefore scope emerging for a creative new co-evolution between top-down innovations in framing conditions and bottom-up innovations in ways that ecosystems are exploited at lower levels of scale. The more bottom-up sustainability initiatives that emerge and the more successful these are the greater become the chances of reinforcing and extending top-down changes in framing conditions and vice-versa.

8 Case Histories from Latin America

Against this backdrop it is useful to review a small number of case histories of bottom-up innovations – in this case chosen from Latin America – to see what lessons and insights they reveal into the governance challenges of sustainable development and how these have been and are being met, including with the help of sustainability scientists. This is important especially to provide inspiration and existence proof that communities facing threats to their livelihoods and security, including environmental change threats, can come together to construct context-specific local solutions that are robust, resilient, fulfilling and equitable and that help immunise them from powerful macro-scale forces that are beyond their immediate control. Such local context-specific solutions have emerged in some instances spontaneously and in other cases have been facilitated or reinforced by policy and market reforms associated with new international policy regimes.

Our first case history concerns a community-based forest governance model that has been developed in the Lachua region of Guatemala. Since 2005 the local community has obtained payments under a PES scheme to protect 3,500 ha of forests and to reforest 2,000 ha of degraded forest areas. Decision making and the management of PES receipts are the responsibilities of democratically-elected community representatives who also develop internal mechanisms to maximize benefits and guarantee an efficient and equitable re-distribution of funds among community members. Parts of the PES receipts are used to financially underpin other sustainable production initiatives, to support a capacity-building programme, and to provide sources of micro-funding. These are aimed at combating local poverty, but also at rebuilding social capital that was seriously damaged in the course of a long civil war. The Lachua community is now moving to access new funding through REDDS and a Forest Incentives Program, PINFOR. Its experience is serving as a model and a pilot for a national 'REDD- Readiness' programme throughout Guatemala.

Our second case history concerns the Río Plátano Biosphere Reserve located in the Moskitia of Honduras. The Río Plátano area faced strong ecosystem degradation and biodiversity loss owing to unsustainable forestry practices and overhunting. These practices were linked to poverty and to conflicts and competition between different ethnic subgroups. To address these challenges local communities came together to agree upon alternative ways of exploiting their shared ecosystems by producing non-timber forest products, including cocoa, ornamental plants, medicines, and ojon oil, which is used for cosmetics. Of special interest is the community-based governance model around ojon oil production and trade. Production is carried out on collective land. The production system relies on an endemic tree species, *Bactris balanoides*, and uses traditional knowledge about ways to extract oil from the tree. This tree species is highly resistant to hurricanes, so this regime of exploitation is also more robust in the face of climate change. The arrangement generates over $1.3 million annually in sales of ojon oil to cosmetics manufacturers. Receipts are distributed among three indigenous federations, two municipalities, and a range of support organizations in the territory, including a local NGO.

Our third case history concerns the Communal Reserve of Tamshiyacu Tahuayo in Perú, an area of fertile upland forest between the Amazon and Yavari rivers on the border of Brazil. The Reserve is now recognised officially and it has been renamed and extended so that it now covers more than 400,000 ha, but it was constituted originally by the actions of a community organization. Local hunters worried by the decline of wildlife and degradation of their forest ecosystem established a long-term alliance with scientists linked to the Wildlife Conservation Society. They jointly engaged in basic research and in the development of management plans for sustainable hunting. This led to the declaration of a communally-managed protected area and to new economic initiatives for eco-tourism. A newly-added activity is carbon sequestration.

Our fourth case concerns a group of 5,000 small coffee growers who have territorial control over the upper watersheds of the Inambari and Tambopata rivers in the Peruvian Amazon. Facing a decline of land productivity, environmental watershed degradation, and low profits from traditional coffee production, the farmers formed a cooperative – the Coffee-grower Cooperative Central of the Sandia Valleys (CECOVASA). At that time coffee growing in the area was focused by competition among growers on maximising the quantity of coffee produced and lowering the economic production costs to individual growers, a regime that contributed to overproduction while also damaging soils and surrounding vegetation and habitat. In 1998, with the help of an international conservation NGO, CECOVASA developed an ambitious programme of farm conversion to shade-grown, organic coffee of high quality and value. The switch away from intensive production has been highly successful. It has delivered higher and more-stable incomes and enabled the region and its growers to develop a reputation in a distinct market segment. CECOVASA has won national and international quality awards for its coffee and for production methods that sustain the ecosystem and the underlying asset base for future production. CECOVASA now has ambitions to extend lessons from its experiences to create a coordinated community of cooperatives able to deliver integrated environmental management of the Tambopata river basin at the full watershed scale, including programmes of land restoration, reforestation, and the development of PES schemes.

Our fifth case is an initiative that is being carried forward jointly by two community-based governance organizations, the Akumal Ecological Centre (CEA) and the Union de Vecinos Akumal (UVA) in association with the University of Quintana Roo in Mexico. Prompted by concerns for increasing risks of extreme weather events – especially hurricanes – arising from climate change, a Climate Change Commission has been established in Quintana Roo and a Climate Change Action Plan has been drafted by the Mexican Government with technical support from a regional University. Adaptations to climate change are being developed in association with two local community-based governance organisations, both of which, in very different ways, are committed to better environmental governance and are themselves closely linked to outside organisations offering scientific expertise and advice. Both of these are associated with the Akumal ecological and tourist enclave located 80 km south of Cancun. CEA represents the permanent resident-tourists and UVA represents local workers who service the complex. Despite evident differences in wealth and incomes, both parties have a shared interest in sustaining the pristine coastal, coral and mangrove ecosystems around the enclave, from which they both benefit, and in finding effective responses to environmental challenges, including climate change. Historically, CEA has monitored the impact of tourist and other developments on regional ecosystems. UVA has meanwhile fought to establish workers' entitlements to land and basic services. Both organizations have now extended their mandates and, supported by sustainability scientists, are working together to develop joint responses to climate change in an action that draws both on ecologies of poverty and ecologies of wealth.

Our last case history concerns the Maquipucuna Foundation and the Choco Andes Conservation Corridor in Ecuador. The Maquipucuna Foundation has had a major influence not only on the sustainable governance of the local ecosystems that it helps manage, but on changing understandings and attitudes to conservation in Ecuador as a whole. The Foundation has raised the profile of the Ecuadorian cloud forest and changed the earlier mindset that conservation represents a loss of economic opportunity and sovereignty. Its first initiative was to establish a 6,000 ha reserve. This provides habitat for around 2,000 vascular plant species and 376 bird species and it supports ecotourism, which is now a main source of income for thousand of families in the region. Its next major initiative was to develop a conservation corridor that provides altitudinal connectivity from the mountains to the sea and regimes for its sustainable exploitation. A regime that incorporates shade-grown cacao in the lowland areas and shade-grown coffee in the highlands is generating good incomes while simultaneously supporting habitat connectivity.

The most recent accomplishment has been to help secure property rights over 44,000 ha for the Esmeraldas afro-ecuadorian community, which is the outcome of a decades-long struggle.

9 Reflections and Final Remarks

Sustainable development is a paradox. On the one hand there is a need for a radical overhaul and restructuring of our current social-ecological systems, which demands change. On the other hand there is a need to secure social, economic, financial and political stability on a continuing basis. To reconcile these conflicting requirements calls for a careful moderation of the pace of change and care in choosing transition pathways so that these are smooth and avoid catastrophic discontinuities and disruptions.[4]

The overall need is to achieve radical change in the global social-ecological system through orderly and evolutionary change processes based on incremental steps that gradually weaken the structures and mechanisms that contribute to unsustainability and replace these with more sustainable structures and mechanisms. This calls for new kinds of multi-level change processes that involve a dynamic interplay and modulation between gradually-introduced top-down changes in framing conditions, which increase local autonomy and give strategic steer, and sets of context-specific innovations that are developed bottom-up and in patchwork fashion, which are enabled in part by such top-down innovations. Bottom-up solutions developed in this way should contribute to improving local sustainability and also reinforce the initial top-down changes and support their further extension.

Our case histories illustrate that this is arising now, for example, through the creation of new markets for ecosystem services. They illustrate also that effective, cooperative and sustainable solutions can come about through self-organisation at the local scale in response to challenges faced and to opportunities that changing frameworks, such as PES opportunities, create. Conditions conducive to such self organisation have been identified already: predictability of higher-level system dynamics, leadership, social capital, common understanding of the social-ecological system and autonomy to make own operational rules, among others (Ostrom 2007). Our case histories illustrate that access to scientific support, both of a technical

[4]This goes some way to explaining why there are no panaceas to the sustainable management of natural resources and ecosystems. Resource degradation commonly has multiple causes rooted in complex interactions between the natural and social components of social-ecological systems. Social-ecological systems are complex, multi-level systems that are not amenable to 'one-size-fits-all', 'silver-bullet', or 'quick-fix' solutions. One-dimensional solutions, unidirectional solutions (top-down or bottom-up), and generalised approaches are all inappropriate. By contrast, sets of carefully chosen changes introduced top-down (especially those that affect markets, which are our most powerful mechanisms of global coordination) may combine synergistically to provide new scope and viability for alternative and more sustainable solutions at local scale to emerge bottom-up.

nature and, more especially, to facilitate processes of social learning, issue reframing, conflict resolution and governance is likely to be another such condition.

The case histories provide inspiration and existence proof that in the face of threats to their livelihoods and security, communities can come together to take advantage even of small windows of opportunity for positive change, including opportunities created around the politics of climate change and biodiversity conservation.[5] They show that, far from acting as 'constraints' on development, international responses to global concerns for climate change, habitat destruction and biodiversity loss in the form of establishing safety boundaries, can increase the degrees of freedom open to local communities over their choices for managing local common pool resources more sustainably, for example by financially or otherwise underpinning sustainable livelihood and sustainable ecosystem exploitation options with payments for ecosystem services (PES).

They also reveal how the concept and politics of climate change can create mobility in forms of governing and social values, which in turn can create opportunity and scope for transformation in the handling of transition to sustainability. Substantial direct non-material benefits to the concerned communities and their members can also be achieved through the governance processes, for example through greater and more inclusive involvement of community members in decision making and by community members taking greater control over their own lives, resources, environment and destiny. Such processes provide routes not only for empowering marginalised groups, but also for satisfying important social and psychological needs, which adds directly to the wellbeing and welfare of communities and individuals. For example, they may serve to win respect for and raise the self-esteem of individuals and groups, such as the poor, ethnic or religious minorities, women and children, who might otherwise be overlooked or excluded, but whose involvement is critical for sustainability.

The challenge of implementing a strategy based around sustainability transitions of the kind just described implies new, additional and different roles for science and scientists in facilitating mediation and modulation between top-down and bottom-up innovations. Especially it suggests a key role in supporting processes of community-based governance of ecosystems and in helping communities to develop and implement bottom-up initiatives that take advantage of opportunities as these arise. These are roles that even go beyond applied problem solving to involve also nurturing mobility in forms of governance and in social values. They are not easy

[5]The case histories describe instances where communities facing threats to common-pool ecosystems and environmental resources upon which they depend have overcome destructive rivalries and have come together to cooperate in developing new and more sustainable regimes of ecosystem management. A shared understanding of the nature and significance of the threats faced is an important factor in the emergence of new governance arrangements and of alternative ways to use and conserve ecosystems. The alternative regimes often involve new management rules, which restrain certain practices. Arrangements for cost and benefit sharing are often integral elements of the governance approach and may include mechanisms for compensating those who lose from changes in management regime and for ensuring an equitable distribution of benefits from the new regimes.

roles for scientists to play or ones that fit well with scientific cultures and conventions or with prevailing organisational and institutional arrangements in science. However, it is through these roles – played out locally in situations and contexts throughout the world over the coming years as communities everywhere come to search for new solutions to the ways they meet their needs and take a greater control over their own destinies and environment – that scientists will probably have most impact on the prospects for sustainable development. Important theoretical concepts developed recently in the emerging field of Transition Theory (Grin et al. 2010), Sustainability Assessment (e.g. Weaver and Rotmans 2006; Rotmans et al. 2008) and Ecosystems Approaches provide tools and methods for scientists to apply in the task.[6]

References

Grin, J., Rotmans, J., & Schot, J. (2010). *Transitions to sustainable development* (418pp). London: Routledge.

International Energy Agency. (2008). *World energy outlook 2008*. Paris: IEA.

Jackson, T. (2009). *Prosperity without growth: Economics for a finite planet* (p. 264). London: Earthscan.

O'Riordan, T. (2008). Some reflections on the conditions for favouring integrated sustainability assessment. *International Journal of Innovation and Sustainable Development, 3*(1/2), 153–162.

Ostrom, E. (2007). A diagnostic approach for going beyond panaceas. *Proceedings of the National Academy of Sciences, 104*(39), 15181–15187.

Rockström, J., Steffen, W., Noone, K., Persson, Å., Chapin, F. S., III, Lambin, E., et al. (2009). Planetary boundaries: Exploring the safe operating space for humanity. *Ecology and Society, 14*(2), 32. [online] URL: http://www.ecologyandsociety.org/vol14/iss2/art32/.

Rotmans, J., Jäger, J., & Weaver, P. M. (2008). Editorial, special issue on integrated sustainability assessment: Concept, process and tools. *International Journal of Innovation and Sustainable Development, 3*(1/2), 1–8.

Steffen, W., Sanderson, A., Tyson, P. D., Jäger, J., Matson, P. M., Moore, B. III, Oldfield, F., Richardson, K., Schnellnhuber, H. J., Turner, B. L. II, & Wasson, R. J. (2004). *Global change and the earth system: A planet under pressure* (336pp). New York: Springer.

United Nations Environment Programme. (2007). *Global environment outlook, GEO-4: Environment for development* (572pp). Nairobi: UNEP, [online] URL: http://www.unep.org/geo/geo4.asp

Weaver, P. M., & Rotmans, J. (2006). Integrated sustainability assessment: What is it, why do it, and how? *International Journal of Innovation and Sustainable Development, 1*(4), 284–303.

Worldwatch Institute. (2008). *State of the world 2008: Innovations for a sustainable economy* (288pp). Washington DC: Worldwatch Institute.

[6]Methodologically, Integrated Sustainability Assessment (ISA) combines three elements: an integrated systems analysis, which seeks to secure broad scope for the assessment; a multi-level and agent-based analytical approach, which seeks to understand multi-level processes that can lead to structural change and transition; and a cyclical, participatory process architecture, which seeks to promote social learning among stakeholders through an empowering dialogue, experimentation and transformative capacity building. ISA offers an approach for stimulating the social and institutional processes that contribute to emergent solutions.

Risks and Opportunities for Sustainability Science in Europe

Jill Jäger

1 Introduction

This chapter explores some of the issues around the topic of "sustainability science". In doing so, it attempts to draw a distinction between the wide variety of disciplinary and interdisciplinary research that can be called "research to support sustainable development" and an approach, here referred to as "sustainability science", that is much more strongly oriented towards the development of strategies and the implementation of measures to deal with problems of unsustainable development. Before discussing the different approaches, however, the chapter examines the need for this kind of research, which arises because of the increasing amount of evidence that despite international agreements and action plans at all scale levels, there has been no success over the past few decades in reconciling human development with the environmental limits of Planet Earth and in securing well-being for all people on this planet now and in the future.

The chapter then discusses a number of reasons why sustainability issues are hard to deal with. First there is the complexity of the problems, with multiple human activities as drivers of change as well as the complex interactions within the Earth System such as those between the atmosphere, the oceans and the land surface. Dealing with the problems involves many different stakeholders, both those whose activities are driving change and those who are affected by environmental and societal changes. Reconciling the perceptions and visions of all of these stakeholders is a major challenge. A further challenge is the immense uncertainty surrounding many aspects of problems of unsustainable development. While research can reduce or eliminate some uncertainties, some of the uncertainties will remain. In particular, uncertainties about how humans might behave in response to given stimuli will remain uncertain.

J. Jäger (✉)
Independent scholar, Vienna, Austria
e-mail: Jill.jager@speed.at

C.C. Jaeger et al. (eds.), *European Research on Sustainable Development*,
DOI 10.1007/978-3-642-19202-9_13, © European Union, 2011
Published by Springer-Verlag Berlin Heidelberg 2011

Given these challenges, the chapter then explores how a "sustainability science approach" could contribute to finding and implementing solutions to persistent problems of unsustainability, in particular through the design of processes of dialogue between all stakeholders, experimentation and learning. Examples are given of networks that are already attempting this kind of work and of the potential of this approach in Europe.

There are, however, also barriers to this kind of work and these are discussed in a further section of this chapter, which suggests that some major changes to the way that science is organized and funded are required for widespread use of sustainability science approaches. The chapter then uses examples from the Conference "Sustainable development: a challenge for European research", several of which are included in the rest of this book, to explore how the challenges and barriers to sustainability science are being dealt with. Much research clearly includes important elements of sustainability science approaches, but there are only a few examples in which the approach is adopted fully.

Over the past decade, sustainability science has advanced in a rather ad hoc manner, with different approaches being tried in different places. Clearly there is a developing need for some consolidation through comparison of similar cases. At the same time, there is a need to address the institutional and other barriers to this kind of work. There are many opportunities to begin processes of transformation towards sustainability with the support of sustainability science. The risks lie in an inability to change the way that science is funded and evaluated, so that the potential of the approach cannot be demonstrated and the ad hoc nature of the endeavour leads to its dismissal.

2 Why Is Transformative Research Needed?

Research over the last two decades has documented that the Earth is undergoing major environmental and socioeconomic changes (see, for example, Steffen et al. 2004). The situation is dramatic, in particular because most of the driving forces of environmental change such as economic growth, consumption levels in industrialised countries, the size of the world population, resource use and energy consumption, continue to increase. In fact, as Steffen et al. (2004) have documented, there has been acceleration in the rate of growth of many of these driving forces since 1950. Population has been growing exponentially since industrialization began. Since 1950 the size of the world economy has increased by more than a factor of 15, inequality in wealth is also increasing and between 1960 and 1994 the ratio of income of the richest 20% to the poorest 20% increased from 30:1 to 78:1, world petroleum consumption has increased by a factor of 3.5 since 1960, and urbanisation increased tenfold in the twentieth century (Jäger 2009a). Since the middle of the last century there have been rapid and profound changes in almost every sphere of human activity. Many of these changes affect the environment and the pressure on the Earth's resources and on the planet's capability to assimilate wastes is intensifying sharply (Steffen et al. 2004).

The acceleration of human activities that affect the environment is already having observable consequences. Climate change, land degradation, deforestation, biodiversity loss and changes of water quality and quantity are prominent examples of global environmental changes. As Steffen et al. (2004) show, almost half of the earth's land surface has been transformed through human activities such as agriculture, urbanization, building of dams, deforestation etc. More nitrogen from the atmosphere is now fixed by the production of fertilizers and burning of fossil fuels than is fixed naturally. More than half of all accessible freshwater is now used by humans. The atmospheric concentrations of several climatically important "greenhouse gases", including carbon dioxide, methane and nitrous oxide, have increased significantly as a result of human activities such as the burning of fossil fuels and intensive agriculture. Coastal wetlands have also been noticeably affected by human activities, in particular through the removal of half of the world's mangrove ecosystems. The oceans have also been significantly affected by human activities, through, for example, depletion of fish stocks, ocean acidification, and various forms of pollution.

While international agreements have been reached to deal with many of these problems, the implementation has not always followed and the problems have been tackled in isolation, rather than recognizing the major interactions between them. As a result, despite agreements reached almost 20 years ago at the UN Conference on Environment and Development in Rio de Janeiro, little has been achieved in putting the planet onto a sustainable track. We are faced by persistent problems of unsustainability resulting from an overexploitation of the planet's resources and ability to absorb wastes. Transformative research is needed so that sustainable pathways can be explored and taken.

3 The Challenge for Research

Several characteristics of persistent problems of unsustainability present serious challenges for scientific research. As Rotmans (2006)[1] has pointed out

> These problems are complex, ill-structured, involve many stakeholders, are surrounded by structural uncertainties, and are hard to manage.

For each of the different problems (climate change, land degradation, biodiversity loss, etc.) or problem sector (agriculture, energy, transport, etc.) the symptoms of unsustainability mask deeper underlying problems in our societal structures and institutions. Thus, as Rotmans (2006) underlines, they cannot be solved in isolation. The complexity arises because of the multiple and interacting drivers of change (e.g. agriculture requires land, water and energy), the interactions within the earth system (e.g. between the atmosphere and the oceans or between climate and vegetation), the interactions between levels of scale (local, regional global), time delays in responses and because of the massive complexities of human consumption

[1]http://www.matisse-project.net/projectcomm/uploads/tx_article/Working_Paper_4_01.pdf

and production systems. The persistence of the problems is because of what Rotmans (2006) refers to as "system failures" –

- *Institutional system failures* (e.g., dominance of institutions that block innovation)
- *Economic system failures* (e.g., inadequate market development or lack of investment capital)
- *Social system failures* (e.g., unchanged behaviour and habits)
- *Ecological system failures* (e.g., dominance of species or ecosystems that threaten biodiversity).

Uncertainty is also an inherent characteristic of persistent problems of unsustainability. There is much discussion in the scientific literature about sources and types of uncertainty (see, for example, references cited by van Asselt (2009)). The latter distinguishes between two sources of uncertainty: variability and lack of knowledge. Sources of variability uncertainty include the inherent variability of natural processes, value diversity as a result of differences of people's world views, behavioural variability (different responses by different people or discrepancies between what people say and what they actually do), societal variability (the unpredictable nature of societal processes) and technological surprises (new developments or unexpected side-effects of technology). Lack of knowledge is partly a result of the above kinds of variability but there are many other sources including : measurement errors, lack of observations or measurements, competing interpretations of available data. For many aspects of persistent problems of unsustainability uncertainty will never be eliminated. Some processes can never be fully predicted or determined.

The complexities and uncertainties, together with the fact that there are multiple stakeholders, mean that normal scientific research projects are ill-equipped to deal with persistent problems of unsustainability. The challenge is further compounded by the need to link knowledge production more closely with action.

4 Meeting the Challenge: Sustainability Science

In October 2000, a small international group of scientists met in Sweden to discuss the emergence of 'sustainability science' (Kates et al. 2001). In response to the kind of challenges discussed briefly in the previous section, it was agreed that approaches are needed that consider the human-environment system as a whole. Because of the need to bridge knowledge and action, the focus of the research must be on a particular place (and the human-environment interactions at that place but also with other places) or a particular sector (again taking into account the interactions with other sectors). It was recognized that fundamental advances would be needed in order to address such issues as the behaviour of complex, self-organizing systems, as well as the responses of the human-environment system to multiple and interacting stresses (Jäger 2009b).

Further discussions on sustainability science (ICSU 2002a) emphasized that research and development (R&D) priorities should be set and implemented so that science and technology contribute to solutions of the most urgent sustainability problems as defined by society, not just by scientists. ICSU (2002a) underlined many of the points raised in the previous paragraph: the focus of on the complex, dynamic interactions between nature and society, the 'place-based' or 'enterprise-based' nature of the work, and the need to consider cross-scale interactions. The bridging between knowledge and action, i.e. research leading to policy formulation and implementation was again emphasized.

In the meantime, a body of research has begun to accumulate that attempts to fulfil these aims. Journals on sustainability science have been established and books have documented some of the research.[2]

In Europe, in particular, sustainability science has evolved towards being strongly implementation-oriented in areas dealing with persistent problems of unsustainability that have a high level of complexity. This implementation orientation presents a number of challenges, in particular because it means that the researchers (or better the practitioners of sustainability science) focus on possible implementation within social, economic and culturally shaped environments. The sustainability scientists, as part of an iterative learning process with selected relevant stakeholders, have to think and act strategically.[3] In this respect sustainability science can be seen as a driver of societal learning and change processes (SLC). This intention to actively contribute to SLC processes distinguishes sustainability science from other approaches. The scientist leaves the "neutral position" as an observer and becomes an active protagonist as part of a process.

Clearly sustainability science is not a "mature discipline" but the discussions and projects over the past decades have begun to clarify its characteristics. It is important to note, however, that the paths taken in Europe, Japan and the USA have been different. While European practitioners have moved towards participatory, iterative processes with an implementation orientation, Japan started with a technology-based approach and has only recently begun to pay more attention to the human dimensions and the USA still prioritizes interdisciplinary research on complex human-environment systems.[4] Despite the differences in approaches,

---

[2]See for example, Moll and Zander (2006), Adger and Jordan (2009), Kasemir et al. (2003), Spangenberg (2008), Ehlers and Krafft (2006). Journals include Sustainability Science (http://www.springer.com/environment/environmental+management/journal/11625), the sustainability science section of the Proceedings of the National Academy of Sciences (http://www.pnas.org/site/misc/sustainability.shtml), Current Opinion in Environmental Sustainability (http://www.elsevier.com/wps/find/P09.cws_home/cosustnews).

[3]The MATISSE project (www.matisse-project.net) provides an example of the design and running of such a strategic process.

[4]This was confirmed by discussions at the International Conference on Sustainability Science in Rome, June 2010. The European approach was discussed at a small workshop in Brussels in October 2009 (http://ec.europa.eu/research/sd/pdf/workshop-2009/report_workshop_sustainability_science_october_2009.pdf#view=fit&pagemode=none). US research was discussed at a workshop in Airlie House, near Washington DC, in November – December 2009 (http://www.nsf.gov/awardsearch/showAward.do?AwardNumber=0955699)

there are clearly some crucial distinctive criteria of sustainability science, which have, for example, been listed by Kasperson (2008) and include the problem-driven nature of the work, the goal of looking at the system as a whole and tackling issues like complexity and uncertainty. The European approaches are further characterized by the focus on designing processes that are open, inclusive and goal-searching and that support learning and change through dialogue, experimentation and joint visions of a sustainable future.

5 Communities of Sustainability Scientists

As outlined in previous paragraphs, there is a growing interest in embracing sustainability issues in research and technology development. In Europe sustainable development has been introduced into a wide spectrum of scientific fields. The sustainable development website of the Directorate-General for Research of the European Commission[5] lists, for example, the following themes in which sustainable development research is carried out: energy, environment, food, agriculture, fisheries, biotechnology, health, information and communication technologies, nanoscience, security, space and transport.

The community actively pursuing sustainability science in Europe (but also globally) is highly diverse and changing. In fact some scholars have pointed to the risk of putting them into one "niche" – the great diversity of people, their backgrounds, expertise and work experience and the multi-, inter-, and trans-disciplinary nature of the institutions indirectly involved is essential and part of the specific strength of sustainability science in Europe (Jäger 2009c). There is no coherent overarching umbrella organization for the various strands of sustainability science work.

The small communities of sustainability scientists that do exist are often oriented towards specific topics e.g. climate change, development, water management, biodiversity, etc. and also technical or economic topics like consumption, production, logistics, and energy. There are few connections between these communities (apart from some individuals). The European Sustainability Science Group (ESSG)[6] is a first step in community building. The individuals in ESSG and the institutions they come from are a good starting point and a fair share of places where sustainability science is done in Europe at present, but the group is too small to fully represent sustainability science. A subset of the ESSG is working together with other colleagues in a working group on the "science-policy interface" of the ESF-COST Forward Look "Responses to Environmental and Societal Challenges for our Unstable Earth (RESCUE)". The goal of RESCUE is to develop a series of key recommendations aiming at improving the development and the impact of the RESCUE-related science community.

[5]http://ec.europa.eu/research/sd/index_en.cfm?pg=fp7-sustainability

[6]www.essg.eu

Other networks active in the area of sustainability science in Europe are listed in Table 1.

6 Opportunities for Sustainability Science in Europe

As highlighted on the sustainable development website of the Directorate General for Research of the EU Commission, the renewed EU Sustainable Development Strategy (EU SDS) adopted in June 2006 assigns an important role to research and development. The Seventh Framework Programme (FP7) for research responds to this challenge with an emphasis on delivering research to support the EU sustainable development objectives. Conversely, many recent major policy documents from the European Commission in areas relevant to sustainable development, ranging from marine policy to energy policy and technology policy, have underlined the importance of research.

EU Member States and Associated Countries have also begun discussions on how research contributes to sustainable development. The different approaches taken in individual countries were discussed at a workshop that took place in June 2007.[7] The participants agreed on a need to follow up on (1) reinforcing the synergies between national and European strategies for putting research at the service of sustainable development, (2) monitoring to what extent the sustainable development potential of FP7 will be translated into reality and (3) improving the role of research in policy making by introducing the idea of knowledge brokerage.

At the conference "Sustainable development: a challenge for European Research", upon which this book is based, one of the sessions discussed a background paper prepared by an expert group on research and development for sustainable development (RD4SD).[8] The RD4SD exercise aimed at discussing how European research can be harnessed for sustainability. The mandate required the expert group to explore the three following questions:

1. To what extent does sustainable development require changes in the way we carry out research?
2. To what extent does sustainable development require changes in the way we elaborate research policies?
3. Which indicators do we need to grasp the contribution of research to sustainable development?

All of these activities – the linking of the Seventh Framework Programme to the renewed EU Sustainable Development Strategy, the activities of Member States and Associated countries and the RD4SD exercise – largely represent a paradigm of

[7]Research for sustainable development – How to enhance connectivity. http://ec.europa.eu/research/sd/pdf/background_info/report_halfman.pdf

[8]The report is available at http://ec.europa.eu/research/sd/pdf/rd4sd/rd4sd_final_report.pdf#view=fit&pagemode=none

Table 1 Sustainability science networks

ESSG (European Sustainability Science Group) www.essg.eu	The European Sustainability Science Group – ESSG – consists of researchers and consultants in the fields of global change research and development research. Present. The overall vision of ESSG is a more implementation-oriented way of carrying out research projects for sustainable development: linking the worlds of science and practice – of knowledge and action.
Td-net (Network for transdisciplinary research) http://www.transdisciplinarity.ch/e/index.php	The network was launched in 2000 by the *Swiss Academic Society for Environmental Research and Ecology*. Since 2008 the *td-net for transdisciplinary research* has been a project of the *Swiss Academies of Arts and Sciences.*
TIAS (The Integrated Assessment Society) http://www.tias.uni-osnabrueck.de/	The Integrated Assessment Society (TIAS) is a not-for-profit-entity created to promote the community of inter-disciplinary and disciplinary scientists, analysts and practitioners who develop and use integrated assessment. The goals of the society are to nurture this community, to promote the development of IA and to encourage its wise application.
ESSP (Earth System Science Partnership) http://www.essp.org/	The ESSP is a partnership for the integrated study of the Earth System, the ways that it is changing, and the implications for global and regional sustainability.
The Resilience Alliance http://www.resalliance.org/1.php	The Resilience Alliance is a research organization comprised of scientists and practitioners from many disciplines who collaborate to explore the dynamics of social-ecological systems. The body of knowledge developed by the RA, encompasses key concepts of resilience, adaptability and transformability and provides a foundation for sustainable development policy and practice.
European Research Network on Sustainability Transitions http://www.ksinetwork.nl/conference2009/	The 1st European Conference on Sustainability Transitions in June 2009 brought together a rapidly growing community of researchers and practitioners interested in broad societal transitions towards sustainability. The common goal is to inform strategies for the governance of sustainability through a better understanding of the dynamics of transitions.
Forum: Science and Innovation for Sustainable Development http://sustainabilityscience.org/	The Forum on Science and Innovation for Sustainable Development is an attempt to outline the burgeoning field. The Forum focuses on the way in which science and innovation can be conducted and applied to meet human needs while preserving the life support systems of the planet. It highlights people and programs that are studying nature-society interactions and applying the resulting knowledge to create a sustainability transition around the world.

"science for sustainable development" in which science is carried out in a traditional mode and contributes to informed decision making. The reference in the 2007 workshop to "knowledge brokerage" provides a first hint of the need for sustainability science. The RD4SD exercise also refers to some elements of sustainability science but does not call for a major shift in the organization and funding of research to respond to the persistent and complex problems of unsustainability.

What could implementation-oriented sustainability science contribute to sustainable development in the EU? The renewed EU sustainable development strategy (SDS) identifies 7 key challenges:

- Climate change and clean energy
- Sustainable transport
- Sustainable consumption and production
- Conservation and management of natural resources
- Public health
- Social inclusion, demography and migration
- Global poverty and sustainable development challenges.

Each of these challenges represents a set of complex issues, broken down into operational objectives, to which sustainability science can contribute through organising iterative processes in which stakeholders (including the research community) develop a common view about the scope of the problem, elaborate a common long-term vision for the future in this problem area and explore the possible pathways to achieve that vision using a variety of scientific tools and methods. Sustainability science can contribute both an approach to dealing with these challenges and support for decisions and implementation. Through structured dialogue processes that are perceived to be credible, salient and legitimate by all participants, sustainability science can make a robust contribution to implementation of the renewed SDS and thus to sustainability transitions.[9]

An even larger contribution of sustainability science would be to design and implement processes that lead to a reconciliation of the sustainable development agenda with the goals of economic growth and employment. This would lead to a reframing of dominant science-policy paradigms of growth and development and replacing them by a more science-based integrative paradigm that looks at human well-being and wealth creation–destruction from a perspective more embedded in the social-ecological system.[10]

Overall, sustainability science could provide examples of good practice of how science and technology can best contribute to finding and implementing sustainable solutions and could help to develop social networks and social capital to improve quality of life and well-being from a global systems and cooperative perspective. Sustainability science could provide some crucial insights for policy making about how to develop new patterns of socio-ecological systems structure formation that are more resilient and less vulnerable to global environmental risks, including potential shocks derived, for example, from resource scarcity, climate change or the emergence of new diseases.

There are two particular areas where sustainability science can make an important contribution: (1) dealing with complexity not by developing single solutions for single problems but considering interdependencies (and trade-offs); (2)

---

[9]For an example of such processes, see the MATISSE project (www.matisse-project.net)

[10]See, for example, http://www.seri.at/index.php?option=com_content&task=view&id=839&Itemid=408

providing a transdisciplinary approach that fosters joint production of solutions in a societal context that makes implementation more effective than other approaches tackling these challenges.

7 Barriers

The problem of unsustainable development was explained very clearly at the United Nations Conference on Environment and Development in Rio de Janeiro in 1992. Since then little progress has been made in solving the problems (UNEP 2007). Sustainability science, as articulated more than a decade ago and implemented in a relatively small number of projects compared to traditional research projects, has a potential to support transitions to sustainability but that potential can only be unleashed, if barriers are overcome.

Underlying several of the barriers is the belief within the scientific establishment (scientists, science policy makers and funders) that taking a strategic approach towards specific implementation is still considered by many to be going beyond the remit of science (Jäger 2009c). Therefore, scientists rarely have a mandate to engage in this kind of work and academic institutions rarely give credit for these "hands-on" strategic processes of engagement. Scientists who do become involved in processes that not only analyse problems and discuss possible solutions but also support both the selection and implementation of measures to deal with the problem are leaving the realm of analysis and assuming the role of an active stakeholder.

In general, current peer-review and project evaluation procedures generally do not support this type of work (see, for example Weaver and Jansen (2004)). Peer review of proposals still looks in most cases for traditional research projects that have a scientific objective with one or more central research questions, a methodology to approach these questions and a list of expected results. Implementation-oriented sustainability science cannot determine a specific objective *ex ante*, because the problem to be dealt with has to be agreed to first with the other stakeholders. Sustainability science is "goal-searching" and not "goal-driven" (see, for example, Weaver and Rotmans (2006)).

Project evaluation of iterative processes that incorporate internal evaluation, learning processes and adaptive management is also problematic. Normal criteria for the success of a project generally start with asking whether the goals of the project have been achieved, which is not possible if the goal or objective is not determined before the project starts. Furthermore, interim external evaluation of iterative sustainability science projects is ill-equipped to deal with the adaptive management explicitly build into the project to allow learning when mistakes have been made.

A substantial barrier to producing a body of experience with sustainability science is the absence of long-term funding to support iterative, participatory processes for a range of geographical contexts and persistent problems of sustainability. The kinds of projects that are necessary for building this body of experience do not fit in the normal mode of a 3- or 4-year project cycle.

The iteration and learning require a longer time and the processes as a whole require a funding commitment for such a period.

As a result of these barriers (although there are a few exceptions[11]) experience so far has been accumulated in an *ad hoc* fashion, with shorter term experiments in various places within a range of contexts and for a number of different issues (climate change, mobility, renewable energy transitions, etc.). The concepts, theories and methods of sustainability science need further development within a coherent framework (Jäger 2009c). While there are advantages in multiple, diverse approaches to finding solutions for persistent problems of unsustainability, current efforts are very diffuse and there is a need to begin a consolidation effort.

Most of the barriers to a major, consolidated effort in sustainability science will not be removed without far-reaching institutional change.[12] The existing institutions that support science and technology in the current governance structure for knowledge require major adjustments in order to improve the links between science, policy and society. Thought still needs to be given to the kinds of institutions that can best support the necessary dialogue and science-practice partnerships to contribute to solutions of sustainability problems. A particular institutional challenge is posed by the need to build partnerships with the business and industry sectors. Institutional changes that enhance capacity building in sustainability science are also required. There are some important opportunities for institutional change in the area of sustainability science. As Tàbara (2009) has argued, two key opportunities for institutional innovation are: (1) the risks of climate change and the struggle to reach international agreements on mitigation and adaptation; and (2) new initiatives to reduce poverty and global resource/environmental degradation inequality/inequity by a supporting a new 'global deal' of North–South (and East–West) cooperation.

One important area of institutional change referred to in the last paragraph is in the area of capacity building. Changes in the educational system to strengthen or even introduce training for sustainability scientists are necessary. Some universities (e.g. Lund, Maastricht, Arizona State, Tokyo) have introduced schools/departments for sustainability science, but the number remains very small compared to the perceived demand for practitioners with the mediation skills and systems approach needed for the processes described above. On the other hand there is no point providing training in such skills, if there are no long-term career perspectives for this kind of work, so attention to providing career incentives is also necessary.

An interesting aspect of capacity building was raised in a panel at the 2005 AAAS annual meeting regarding the lack of opportunities for young scientists to engage in and learn from work that directly links knowledge with action in the area of sustainable development. As Clark (2005) reports, the panel consisted of half a dozen young environmental scholars and development activists from China, India

[11]Probably best exceptions are transitions research networks in the Netherlands detail and at least as a model the LTSER expand

[12]A conclusion also reached by the working group on the Science-Policy Interface of the ESF RESCUE project (www.esf.org/rescue)

and Brazil. Their discussions emphasised the need for more recognition by the scientific community of the value of problem-driven work and more support by society in undertaking such work. One very interesting proposal that was generated through the panel discussion is the recognition and support of volunteer efforts by scientists to work 'in the trenches' on pressing problems. In other professions, this kind of volunteer work has long been supported (e.g. in the legal and medical professions). This would contribute to capacity building and to the accumulation of experience. As Clark (2005) points out, setting up a successful programme of scientist volunteers for sustainability would require, above all else, 'that the scientific community and its gatekeepers formally acknowledge the importance of such volunteer work in professional careers'.

8 Meeting the Challenges and Removing Barriers

Using examples from the Conference "Sustainable development: a challenge for European research" (Brussels, 26–28 May 2009) organised by the Research DG of the European Commission, this section explores how the challenges and barriers discussed above are being dealt with. Based on material discussed in previous sections, some important elements of what can be called "sustainability science" are:

- Taking an integrative view of the human-environment system;
- Using a participatory approach;
- Developing a common vision of the future and exploring possible pathways;
- Discussing trade-offs between pathways;
- Linking across scale levels;
- Integrating different forms of knowledge; and
- Fostering learning.

8.1 Taking an Integrative View of the Human-Environment System

Probably the best examples of this element of sustainability science are the papers that won awards at the conference and are included in this volume. Topics such as climate change, energy scarcity and water resources management clearly require an integrative view of the human-environment system since these problem areas arise because of the interaction between human activities and natural systems. Importantly, the papers show that the integration requires methodologies and approaches that differ from traditional research approaches, often involving stakeholders as discussed in the next paragraph.

8.2 Using a Participatory Approach

In recent years, there has been increasing recognition of the value of participatory approaches in dealing with persistent problems of unsustainability (see, for example, Siebenhüner 2004; van de Kerkhof and Wieczorek 2005; Whitmarsh et al. 2009). Even in the conference session largely oriented to the natural sciences ("Interactions and feedbacks between ecosystems and climate change") the presentation on the impact of ocean acidification on marine organisms and ecosystems indicated the inclusion of a "Reference User Group". Participation also played an important role in projects on sustainable cities and the session entitled "Yes, We Did" pointed out that dissemination, a traditional component of research projects, is moving towards co-creation of research outputs. That is, participation of stakeholders is expanding to include the design and use of research results. One session at the conference also focussed on the contributions of civil society organizations to research for sustainable development, pointing out that in addition to technological solutions there must be changes in mindsets but also concluding that civil society organizations have not been much involved in this kind of research so far.

8.3 Developing a Common Vision of the Future and Exploring Possible Pathways

In many of the implementation-oriented processes tackling sustainability issues, it has been found important to include a step in which all participants develop a joint vision (see, for example, Weaver and Rotmans 2006). This helps the participants to become much more innovative in thinking about possible solutions to problems of unsustainability. Developing a vision is not practiced in a large number of projects, but it was demonstrated in projects with a transitions-research approach (see, Chapter 7 in this volume) and in papers from the spatial planning perspective.

8.4 Discussing Trade-offs Between Pathways

Again, this element of sustainability science is standard practice in the transitions-research community and not common in other research projects. An example during the conference was provided by the paper on a transition to sustainable materials management in Flanders.

8.5 Linking Across Scale Levels

Making the linkage between human-environment interactions at the local level and processes at the regional and global levels presents numerous methodological challenges but is essential when sustainability is being assessed. Similarly, linkages between the near-term and the long-term future also have to be considered, not only because of time-lags in the system but also given the possibilities of "tipping points" or thresholds. Linking across geographical scale was exemplified in the work on sustainable primary health care in the session on enhancing global sustainability through international cooperation. Linking across temporal scale was not covered in any detail during the conference.

8.6 Integrating Different Forms of Knowledge

The recognition that scientists are not the only people who can contribute knowledge in implementation-oriented research is, of course, linked to the call for participatory processes, as discussed above. In particular, for solutions-based research, traditional or indigenous knowledge (ICSU 2002b) could play a significant role. The sessions and papers at the conference did show some evidence of the use of different sources of knowledge. In the session on international cooperation the use of indigenous knowledge in Arctic research was discussed. In the session on economics, employment, behaviour and territorial dynamics there was discussion on improving the interface between quantitative and qualitative discourses, while the session "New Imaginings" discussed systems of knowledge governance.

8.7 Fostering Learning

This is one of the central elements of sustainability science. The design of processes that support learning by all stakeholders, including the scientific community, about the causes and consequences of, as well as possible responses to, persistent problems of unsustainability is a necessary part of implementation-oriented research. The learning both about the perspectives on the issue of other actors as well as about the process itself and possible improvements in the next iteration contribute to effective processes (see, for example, Tuinstra et al. 2008).

At the conference the topic of "learning" was raised in several diverse sessions. The session on "Cash and Theory" noted the importance of changing awareness of Chief Executive Officers and employees. Several of the winning papers in this volume also include discussions of learning. Importantly, in the session "Yes, We Did" participants noted the need to provide space for reflexivity in projects and to support the learning of researchers by including a step (or, steps) of self-conscious evaluation within the research process.

8.8 Linking knowledge to Action

One of the dominant themes throughout the conference was the need to link knowledge with action. This was often posed as a question: "What do we need to do to make better linkages?" Or, in the "Fix It" session, "What do we need to do to get widespread implementation of a radical innovation?" Clearly the above seven elements are extremely important in improving the linkages between knowledge and action, as shown in other research and analysis (see, for example, Farrell and Jäger 2005; Cash et al. 2003). Overall, what they point to is the central significance of the design of processes and it was even noted that the researcher can benefit as much from the process as from the final product.

9 Reflection

A number of characteristics distinguish sustainability science from other research endeavours. These have been used in the previous section to examine the content of the conference on research for sustainable development held in Brussels in 2009. We find that each of the characteristics was present in papers presented in the conference, but not all presentations encompassed all of the ingredients of sustainability science. In fact, most of the presentations did not have all of the ingredients. Most of the research presented is certainly "science in support of sustainable development" but does not use the iterative, participatory and implementation-oriented approaches of sustainability science (at least as at is practiced in Europe). To further the use of sustainability science approaches, the barriers discussed in Sect. 7 must be removed.

First, there is a need for more funding mechanisms for the goal-seeking, iterative and integrative approaches to address the complex issues of sustainability. Second, and related to the first need, the review mechanisms for proposals and projects must be modified to deal with the special characteristics discussed in the previous section. Evaluation criteria are required that allow for a continued learning process with stakeholders. Finally, there is a need for improved incentive structures for scholars who wish to engage in implementation-oriented work, where credit could be as much for designing an effective process as for concrete results in the form of a scientific publication.

As noted above, sustainability science is not a mature discipline with shared conceptual and theoretical components. As shown by the other contributions to this book and the discussion above, we find multiple sciences addressing a common theme – the reconciliation of societies' development goals with the planet's environmental limits over the long term (see Clark and Dickson 2003). Sustainability science is heterogeneous in scope and practice. Over the coming years, considerable effort will be needed to begin a consolidation effort. With the removal of barriers it should be possible to demonstrate the huge potential of this approach to stimulate transitions to sustainability.

References

Adger, W. N., & Jordan, A. (Eds.). (2009). *Governing sustainability*. Cambridge: Cambridge University Press.

Cash, D. W., Clark, W. C., Alcock, F., Dickson, N. M., Eckley, N., Guston, D. H., et al. (2003). Knowledge systems for sustainable development. *Proceedings of the National Academy of Sciences of the United States of America, 100*, 8086–8091.

Clark, W. C. (2005). A new social contract for science? *Environment, 47*(3), 1.

Clark, W. C., & Dickson, N. (2003). Sustainability science: The emerging research program. *Proceedings of the National Academy of Sciences of the United States of America, 100*(14), 8059–8061.

Ehlers, E. A., & Krafft, T. (Eds.). (2006). *Earth system science in the anthropocene*. Heidelberg: Springer.

Farrell, A. E., & Jäger, J. (Eds.). (2005). *Assessments of regional and global environmental risks: Designing processes for the effective use of science in decision-making*. Washington, DC: Resources for the Future Press.

ICSU (International Council for Scientific Unions). (2002a). *Science and technology for sustainable development*. ICSU Series on Science for Sustainable Development, No. 9. Paris: International Council for Scientific Unions.

ICSU. (2002b). *Traditional knowledge and sustainable development*. ICSU Series on Science for Sustainable Development No. 4 (p. 24). Paris: International Council for Science.

Jäger, J. (2009a). *Our planet: How much more can earth take?* London: Haus Publishing.

Jäger, J. (2009b). The governance of science for sustainability. In W. Neil Adger & Jordan Andrew (Eds.), *Governing sustainability*. Cambridge: Cambridge University Press.

Jäger, J. (2009c). *Sustainability science in Europe*. Paper prepared for DG Research, European Commission, Brussels. Available at: http://ec.europa.eu/research/sd/pdf/workshop-2009/background_paper_sust_science_workshop_october_2009.pdf#view=fit&pagemode=none (accessed May 28 2011).

Kasemir, B., Jäger, J., Jaeger, C. C., & Gardner, M. T. (2003). *Public participation in sustainability science*. Cambridge: Cambridge University Press.

Kasperson, R. E. (2008, October 29). *Closing the science-practice gap and the role of sustainability science*. Colloquium given at the Institute of Environmental Science and Technology, Autonomous University of Barcelona,Barcelona.

Kates, R. W., Clark, W. C., Corell, R., Hall, J. M., Jaeger, C. C., Lowe, I., et al. (2001). Sustainability science. *Science, 292*, 641–642.

Moll, Peter, & Zander, Ute. (2006). *Managing the interface. From knowledge to action in global change and sustainability science*. Munich: oekom.

Rotmans, J. (2006). *Tools for integrated sustainability assessment: A two-track approach*. Working Paper 4, MATISSE Project. Available at http://www.matisse-project.net/projectcomm/uploads/tx_article/Working_Paper_4_01.pdf (accessed May 28 2011).

Siebenhüner, B. (2004). Social learning and sustainability science: Which role can stakeholder participation play? *International Journal of Sustainable Development, 7*, 146–163.

Spangenberg, J. H. (Ed.). (2008). *Sustainable development – past conflicts and future challenges*. Münster: Verlag Westphälisches Dampfboot.

Steffen, W., Sanderson, A., Tyson, P. D., Jäger, J., Matson, P. A., Moore, B., III, et al. (Eds.). (2004). *Global change and the earth system: A planet under pressure*. Berlin: Springer.

Tàbara, J. D. (2009, July 10). *What is sustainability science and what it is not*. Speech given at the Catalan Polytechnic University, Barcelona.

Tuinstra, W., Jager, J., & Weaver, P. M. (2008). Learning and evaluation in integrated sustainability assessment. *International Journal of Innovation and Sustainable Development, 3*(1/2), 128–152.

UNEP. (2007). *Global environmental outlook*. Nairobi: United Nations Environment Programme.

van Asselt, M. B. A. (2009). Uncertainty management in integrated regional assessment. In C. G. Knight & J. Jäger (Eds.), *Integrated regional assessment of global climate change*. Cambridge: Cambridge University Press.

van de Kerkhof, M., & Wieczorek, A. (2005). Learning and stakeholder participation in transition processes towards sustainability: Methodological considerations. *Technological Forecasting & Social Change, 72*, 733–747.

Weaver, P. M., & Jansen, J. L. (2004, July). *Defining and evaluating "science for sustainability"*. Paper presented at the international conference on sustainability engineering and science, Auckland. [Online] Available from: http://www.nzsses.auckland.ac.nz/conference/2004/Session5/63%20Weaver.pdf (accessed May 28 2011).

Weaver, P. M., & Rotmans, J. (2006). Integrated sustainability assessment: What is it, why do it and how? *International Journal of Innovation and Sustainable Development, 1*(4), 284–303.

Whitmarsh, L., Swartling, Å., & Jäger, J. (2009). Participation of experts and non-experts in a sustainability assessment of mobility. *Environmental Policy & Governance, 19*, 232–250.

Concluding Remarks

Carlo C. Jaeger and J. David Tàbara

"The simple fact that physicists split the atom without any hesitations the very moment they knew how to do it, although they realized full well the enormous destructive potentialities of their operation, demonstrates that the scientist *qua* scientist does not even care about the survival of the human race on earth or, for that matter, about the survival of the planet itself." These bitter words from Hannah Arendt's classical essay on "The conquest of space and the stature of man" remind us that the relation between science and sustainable development – a development surely inspired by "care about the survival of the human race on earth or, for that matter, about the survival of the planet itself" – is far from trivial.

Arendt liked science, but in an age of nuclear weapons and intercontinental missiles she saw the spirit of scientific inquiry confronted with a far-reaching crisis. Edmund Husserl, one of her teachers, had diagnosed this as the crisis not only of science, but of what he called European humanity. He wrote: "Spiritually Europe has a birthplace. By this I do not mean a geographical place, in some one land, though this too is true. I refer, rather, to a spiritual birthplace in a nation or in certain men or groups of men belonging to this nation. It is the ancient Greek nation in the seventh and sixth centuries B.C. In it there grows up a new kind of attitude of individuals toward their environing world. Consequent upon this emerges a completely new type of spiritual structure, rapidly growing into a systematically rounded cultural form that the Greeks called philosophy."

Philosophy, so understood, is vital for the existence of a political community – and the other way round. As Arendt put it: "The *polis*, properly speaking, is not the

C.C. Jaeger (✉)
Potsdam Institute for Climate Impact Research, Head of research domain Transdisciplinary Concepts & Methods, Telegrafenberg A 31, 14473 Potsdam, Germany
e-mail: carlo.jaeger@pik-potsdam.de

J.D. Tàbara
Institute of Environmental Science and Technology, Autonomous University of Barcelona, Campus UAB. E-08193 Cerdanyola del Vallès, Barcelona, Catalonia, Spain
e-mail: Joandavid.tabara@uab.cat

C.C. Jaeger et al. (eds.), *European Research on Sustainable Development*,
DOI 10.1007/978-3-642-19202-9_14, © European Union, 2011
Published by Springer-Verlag Berlin Heidelberg 2011

city-state in its physical location; it is the organization of the people as it arises out of acting and speaking together". Power, in her view, is not the ability to impose one's will onto others, but the capacity of people to jointly act while engaging in reasoned debates about their actions. Violence, in turn, arises where power is found wanting, where the capacity to act together is absent to the degree that some people feel compelled to impose their will on others by destroying them. In this perspective, the extent to which science has come to be a means to enhance instruments of violence to levels hitherto unthinkable is indeed sign of a crisis of the sciences, and of the European heritage that made them possible. And scientific inquiries about issues of sustainable development are not simply excursions of scientific curiosity into a new field, but an opportunity to address that crisis.

Both volumes have grown out of the conference "Sustainable Development: A Challenge for European Research", organised in Brussels, May 26–28, 2009, by the European Directorate General for Research in cooperation with the Czech Presidency of the EU. The conference was organized in the spirit of a manifesto by the Scientific Committee reprinted in both volumes. The conference became a high-level gathering of researchers and practitioners, and it helped to crystallize the research community committed to perform research in a perspective of sustainable development. Meanwhile, the conversation is developing further. Therefore, these two volumes are not restricted to papers presented there, but rather combine papers from the conference with other contributions aiming at the same goal: to develop research in a perspective of sustainable development, jointly shaping our common future.

In this spirit, the first of the two volumes is advocating transformative research – transformative, that is, of both the human practices that are currently jeopardizing our common future and of science itself. If this transformation is to succeed, Western science must become both more ambitious and more modest. More ambitious by becoming truly global, by engaging in new ways with non-Western cultures, more modest by becoming one of many voices in the conversation of humankind, on a par with philosophy, but also with religion, with the arts, with the oral traditions rooted in everyday life. This book has argued that sustainable development requires a massive transformation both in the procedures as in the content of research. It also defended that a main source for social and scientific innovation lies in placing sustainable development in the core of the research agenda. Therefore, a main goal of this collection of essays has been to provide the reader with a number of powerful concepts and ideas on how to frame science and engage the public in research activities in a mode which can contribute in a full transformative manner to sustainable development.

Sustainable development does not only call for radical changes in the configuration of social-ecological systems in general but most noticeably for transformations in the core values and worldviews that drive individual actions and organisations. Science can contribute to such changes, but only if such challenges are addressed in an open, exploratory and learning mode. New processes of researching and constructing research partnerships are required together with a new generation of scientists aware of the type of predicament humankind now confronts. A recurrent theme in this book has been the claim that research for sustainable development demands new roles and new types of relationships scientists maintain not only with the 'objects' of their research, but most importantly with the subjects their potential results are eventually to be developed and applied. Sustainability scientists require a distinct type of professional competences – facilitation skills, systems thinking, ethical reasoning, anticipation and abilities to build strategic partnerships, among others. That is, a series of competences that include the ability to listen, to engage with stakeholders, and to respond to needs and values of the particular contexts in which they work, skills which are not often encouraged enough under most of the mainstream educational careers and programmes.

The recognition of these new professional competences for sustainability scientists may also lead to some misunderstandings about its implications for the production of scientific knowledge. In this new situation, some may argue that sustainability scientists cannot produce 'objective' knowledge, and that their insights and processes cannot be extrapolated or can be valid in other contexts beyond those in which they work. Mostly, because they operate under uniquely distinct value and priority systems, although this would be the same of saying that sustainability science is simply not possible. However, multiple epistemologies and practices are valid in the science for sustainability and in fact, the engagement of scientists with stakeholders of different social-ecological systems is a prerequisite for the accurate understanding of the underlying system dynamics – most of which follow mainly a moral structure and content. Rather than a hindrance for the production of a type of 'objective' knowledge, the existence of a diversity of intelligences and of ethical judgements adapted and responding to local conditions and needs is a prerequisite for a science which aims at improving sustainability in a non-teleological way.

Thus, science for sustainability is not only driven by curiosity. Its emphasis remains in searching and implementing alternative pathways of development more robust with the social-ecological contexts in which the new professionals can operate. This transformational ambition of sustainability science is of particular interest for the creation of institutions explicitly aimed at dealing with large scale interrelated risks and problems derived from global environmental change. In the case of emerging technologies, for instance, synthetic biology is not an inexorable future or there is not a single future in the development of this technology. The possible pathways are multiple, each one generating its own system dependencies, and for each one the role of policy and of the public as well as the benefits and potential risks for society and for global ecosystems may vary strongly. To what extent new technologies will create new path dependencies and new lock-in situations in undemocratic and risky ways is something that a well structured sustainability science could address – e.g., to anticipate the more negative unwanted consequences of its widespread implementation.

In short, there are no simple solutions to complex challenges of unsustainability. Any progress in sustainability will require serious philosophical and ethical considerations. However, the fact that sustainable development issues are complex and require moral judgements and responsible choices between multiple trade-offs, does not entail that there are no feasible options for substantive and decisive action. It is the responsibility of scientific imagination and of scientists to engage in their urgent search and implementation. In this book, we have taken the stance that positive alternative pathways of development, more suited to the requirements of sustainability, can be charted, but this will only happen if science also contributes actively and ethically in this endeavour. We already know a lot about the causes of global environmental change and unsustainability as well as of its possible ways to alleviate its most destructive effects. Progress is possible, although never under a single, predetermined, or lineal form. New approaches to integrate knowledge and ethics, to build constituencies of action, and to redirect both the means and the goals for scientific innovation in line with sustainability and build alternative modes of development can and need to be successfully implemented. And we must do that now.

Index

C.C. Jaeger et al. (eds.), *European Research on Sustainable Development*,
DOI 10.1007/978-3-642-19202-9, © European Union, 2011
Published by Springer-Verlag Berlin Heidelberg 2011

MIX
Papier aus verantwortungsvollen Quellen
Paper from responsible sources
FSC® C105338

If you have any concerns about our products,
you can contact us on
ProductSafety@springernature.com

In case Publisher is established outside the EU,
the EU authorized representative is:
Springer Nature Customer Service Center GmbH
Europaplatz 3, 69115 Heidelberg, Germany

Printed by Libri Plureos GmbH
in Hamburg, Germany